Colloids and Interfaces with Surfactants and Polymers – An Introduction

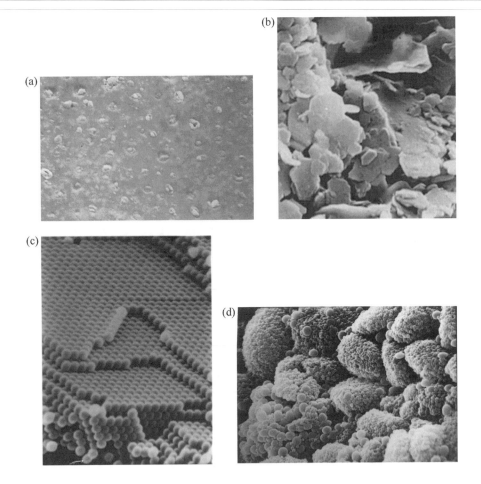

Scanning electron micrographs of some colloidal particles:

(a) A film of printing ink showing well-dispersed titanium dioxide pigment (3400×).
(b) Sodium kaolinite (china clay) particles in an open 'card-house' aggregate (9100×).
(c) Dried polystyrene latex showing how the order in a concentrated dispersion is maintained on drying (9100×).
(d) Coagulated bimodal latex mixture after shear processing showing densely packed, uniform aggregates (700×). Reproduced with permission from 'Modern Aspects of Colloidal Dispersions' edited by Ottewill and Rennie. Copyright (1998) Kluwer Academic Publishers.

Colloids and Interfaces with Surfactants and Polymers – An Introduction

Jim W. Goodwin
Interfacial Dynamics Corporation
Portland, Oregon, USA

John Wiley & Sons, Ltd

Other Wiley Editorial Offices

John Wiley & Sons Inc., 111 River Street,
Hoboken, NJ 07030, USA

Jossey-Bass, 989 Market Street,
San Francisco, CA 94103-1741, USA

Wiley-VCH Verlag GmbH, Boschstr. 12,
D-69469 Weinheim, Germany

John Wiley & Sons Australia Ltd, 33 Park Road,
Milton, Queensland 4064, Australia

John Wiley & Sons (Asia) Pte Ltd, 2 Clementi Loop #02-01,
Jin Xing Distripark, Singapore 129809

John Wiley & Sons Canada Ltd, 22 Worcester Road,
Etobicoke, Ontario, Canada M9W 1L1

Wiley also publishes its books in a variety of electronic formats. Some content that appears in print may not be
available in electronic books.

Library of Congress Cataloging-in-Publication Data

Goodwin, James W. (James William)
   Colloids and interfaces with surfactants and polymers: an introduction / Jim Goodwin.
      p. cm.
Includes bibliographical references and index.
   ISBN 0-470-84142-7 (Cloth : alk. paper) – ISBN 0-470-84143-5 (Paper : alk. paper)
   1. Colloids. 2. Surface active agents. 3. Surface chemistry. 4. Polymers. I. Title.
   QD549.G57 2004
   541′.345–dc22   2003016576

British Library Cataloguing in Publication Data

A catalogue record for this book is available from the British Library

ISBN: 0 470 84142 7 (Cloth)
      0 470 84143 5 (Paperback)

Typeset in 10/12 pt Times by Kolam Information Services Pvt. Ltd, Pondicherry, India
Printed and bound in Great Britain by TJ International, Padstow, Cornwall
This book is printed on acid-free paper responsibly manufactured from sustainable forestry in which at least
two trees are planted for each one used for paper production.

Contents

Preface

We take for granted the behaviour of colloidal systems from a very early age and as we are formerly taught the various scientific disciplines, the nature of colloids is rarely mentioned. This is surprising as it impacts on so many products that we use everyday. The processing and delivery of the correct properties is often dependent on the material being in the colloid state and yet few courses cover the subject in depth, if at all. As a result, many scientific workers have to acquire their knowledge piecemeal while working on other problems. An introductory text is what is most used in this situation. Specialist texts are often left on the shelf as we search for attempts to solve a particular problem. This present text covers a little more ground than some of the older introductory volumes that are still available but I have attempted to help the reader in the more complicated regions by providing a strategy for any calculations. Some derivations are outside the scope of an introduction, but, where they are straightforward, they are included in order to help readers gain as much insight as possible. None of the mathematics is at a high level.

The references are mainly to specialist volumes in the area. This is not to discourage the reading of the original work but it is because the general volumes are more readily available in many libraries and often help to put the work in context. There will inevitably be some areas that are neglected, as with any introductory volume. This is in part due to what areas are perceived to be currently most generally needed and the composition of this volume has been governed by the questions which are currently asked when discussing problems in industry.

I would like to acknowledge with gratitude the help and encouragement of all of my previous colleagues in the Department of Physical Chemistry at the University of Bristol, UK, especially Professors Ron Ottewill and Brian Vincent, the past and current holders of the Leverhulme Chair in that Department and also to Dr Paul Reynolds, the manager of the Bristol Colloid Centre.

Jim Goodwin
Portland, Oregon, 2003

Chapter 1

The Nature of Colloids

1 INTRODUCTION

Colloids are all about size. They consist of at least two phases and the dimension of the dispersed phase has traditionally been considered to be in the sub-microscopic region but greater than the atomic size range. That is within the range 1 nm to 1 μm. The term 'colloid' was coined for 'glue-like' materials which appeared to consist of only one phase when viewed under the microscopes of the day. Of course, now we are able to see much smaller particles with the advance of microscopy. However, the size range is still taken to be the same although 10 μm would be a more appropriate upper limit as the unique behaviour of colloidal particles can still be observed with particle dimensions greater than 1 μm.

The particle size is similar to the range of the forces that exist between the particles and the timescale of the diffusive motion of the particles is similar to that at which we are aware of changes. These two factors, as we shall see later in this volume, are the key to understanding why so many colloidal systems have interesting behaviour and textures. Typically, the range of the interparticle forces is 0.1 to 0.5 μm whether they are forces of attraction between the particles or forces of repulsion. When we look at a colloidal sol in the microscope, we observe the particles to move around with a random motion. This is known as *Brownian motion* as it was recorded by the botanist Brown while studying a suspension of pollen grains in the microscope. The cause of this motion is, in turn, the motion of the molecules making up the suspending fluid. All of the atoms or molecules are in random or thermal motion and at any given instant the local concentration of a small volume element of the fluid will be either higher or lower than the global average concentration. The thermal motion of the colloidal particles will tend to be in the direction

Colloids and Interfaces with Surfactants and Polymers – An Introduction   J. W. Goodwin
© 2004 John Wiley & Sons, Ltd   ISBN: 0-470-84142-7 (HB)   ISBN: 0-470-84143-5 (PB)

of the lower molecular densities. As these fluctuate in a random manner, so does the directional motion of the colloidal particles and the velocity is governed by the hydrodynamic drag. We know that diffusion tends to be away from high concentrations to low concentrations so that if we have a high concentration of particles then there will be a directional drift away from this region. Now for a sphere, the *Stokes drag factor, s*, is a function of the radius of the sphere, a, and the viscosity of the fluid, η, as follows:

$$s_v = 6\pi\eta a \qquad (1.1)$$

The motion is random as we have already noted and the net velocity, v, is the average distance moved, $\bar{x}$, in the time interval t, namely:

$$v = \bar{x}/t \qquad (1.2)$$

The work that has been done in moving a particle is simply the hydrodynamic force, $f_v = vs_v$, multiplied by the average displacement $\bar{x}$. The thermal energy available for this motion is $k_B T$ where T is the absolute temperature and k_B is the Boltzmann constant. Hence we can write:

$$k_B T = \bar{x}vf_v \qquad (1.3)$$

Substituting for v and f_v and rearranging:

$$D = \frac{\bar{x}^2}{t} = \frac{k_B T}{6\pi\eta a} \qquad (1.4)$$

Equation (1.4) is the Stokes–Einstein equation for the diffusion coefficient, D, and has units of $m^2\,s^{-1}$. We can define a characteristic timescale for this diffusive motion if we calculate the time it takes for a particle to diffuse a distance equal to the particle radius. This is achieved by a straightforward substitution of a for $\bar{x}$ in Equation (1.4), as follows:

$$t = \frac{6\pi\eta a^3}{k_B T} \qquad (1.5)$$

This is known as the Einstein–Smoluchowski equation. For an isolated particle in water at 20 °C with a diameter of 1 μm, it takes about 0.5 s to diffuse one radius. When the colloidal dispersion becomes concentrated, the interactions with the neighbouring particles (hydrodynamic, electrostatic if the particles are charged, or simply van der Waals' forces) will slow the movement down. The timescale of our perception is approximately 1 ms to 1 ks and

Table 1.1. Types of colloidal dispersions

Phase	Gas (bubbles)	Liquid (droplets)	Solid (particles)
Gas	Molecular solution	Liquid aerosol (mist)	Solid aerosol (smoke)
Liquid	Foam (shampoo)	Emulsion (mayonnaise)	Sol (ink)
Solid	Solid foam (packaging)	Solid emulsion (butter)	Solid sol (stained glass)

so we should expect to observe interesting temporal behaviour with colloidal systems. We will re-visit this point later in this volume.

When we consider the number of possible phase combinations of our heterophase systems we find that there should be eight different possibilities. This is illustrated in Table 1.1 where either phase could be a gas, a solid or a liquid. Two gas phases will mix on a molecular level and do not form a colloidal system. Each of the other combinations results in systems with which we are familiar.

Gas bubbles and liquid droplets are spherical due to the surface tension forces unless the phase volume is very high. Solid particles may be spherical but are often non-spherical. The shape is a function of the history of the formation. Opals are an example of a solid sol with spherical silica particles in an aqueous silicate matrix. The silica particles are amorphous silica, and the distribution of sizes of the particles is narrow and the particles form a face-centred cubic array. It is diffraction of light by this highly regular structure which gives the characteristic colours. Colloidal dispersions in which the standard deviation on the mean size is less than 10% of the mean are usually considered to be 'monodisperse'. If the particle size distribution is broader than this, the dispersion is considered to be 'polydisperse'. Although this cut-off appears arbitrary, monodisperse systems have the ability to form colloidal crystals while polydisperse systems do not. Bi-modal systems can also form crystalline structures if the size ratio is suitable. When the particles are formed by a crystallization process, other shapes are found. Silver chloride can be produced as a colloidal dispersion in water as monodisperse cubes. Hematite can form as ellipsoidal plates. Clays are naturally occurring aluminosilicates that usually form plates. Kaolinite particles ('china clay') are hexagonal plates with an axial ratio of $\sim 10:1$. Montmorillonite particles can have much greater axial ratios and under the right conditions can be dispersed as crystals of one or two unit layers thick. Attapulgite has a lath shape and longer rod-like structures can be seen with crysotile asbestos. These shaped particles show colloidal behaviour when the size is within the colloid range. For spheres or cubes, we have a three-dimensional colloidal size, with rods this is reduced to two dimensions, while for plates only one dimension needs to be in the appropriate size range. This last case may seem strange but soap films are a good example of a system with two dimensions well within the macroscopic size range but with the third in the colloidal range and being governed by colloidal forces.

This last example of a colloidal system brings into focus systems other than particles that have common ground with particulate colloids. Surface active molecules or surfactants, such as soaps, detergents, lipids etc., can self-assemble to form multimolecular aggregates of colloidal size and show the effects of colloidal forces in addition to their individual phase behaviour.

2 COLLOIDS IN ACTION

It will serve as a useful illustration to take some examples of colloidal systems and discuss why the colloidal state is used, what are the important aspects and what characterization is desirable. Although each colloidal material appears to be very different from others, there are frequently generic aspects and so we can learn from solutions developed for quite disparate systems.

2.1 Decorative Paint

The function of this type of coating is twofold. First, it is intended to protect the surface from damage from environmental conditions. Secondly, it is intended to cover marks and produce an attractive colour. By choosing a colloidal system we are able to manufacture and apply this very simply. A polymer film provides the surface protection. Synthesizing the polymer as colloidal particles dispersed in water can efficiently produce this. This material is known as a *latex* and is manufactured by the emulsion polymerization of vinyl monomers. The latter are dispersed as an emulsion using surface active materials (*surfactants*) which adsorb at the surface of the droplets and prevent them from coalescing. Once the polymerization reaction is initiated, the size and stability of the subsequent particles is also controlled by the surfactants. The advantages of using this colloidal synthetic route is excellent heat and mass transfer and simple handling of the product which can easily be pumped out of the reactor and into storage tanks. Here we have to understand how the surfactants adsorb onto different organic phases and operate at different temperatures.

The covering power of the film is provided by a white pigment and the colour by tinting with coloured pigments. Light scattered from the white pigment particles (usually titanium dioxide) hides the underlying surface. The particles must be fine enough to give a smooth film but not too fine or insufficient light will be scattered – 200 nm is about the optimum size. To manufacture this, we must understand the control of crystal growth and the subsequent drying process to ensure easy redispersion of the dry powder down to the sub-micron level. The surface of the titanium dioxide is usually covered by a layer of alumina or silica to reduce catalytic breakdown of the polymer film when exposed to sunlight. The dispersion of dry powders in liquids requires surfactants and energy. Here, we have to understand how particles scatter light, the separation of colloidal particles and the 'wetting-out' of dry powders and their subsequent redispersion. Thus, this means how surfactants control the wetting of surfaces and how shear forces break up aggregates. The coloured pigments may be organic and therefore require different surfactant systems and so we may put together a system with three different surfactant materials and there will be ample opportunity for exchange at the various interfaces.

The final aspect of our paint is the application. At this point, the sedimentation of the pigment must be controlled and the viscosity has to be such that the wet film thickness is sufficient to give good hiding power. In addition, the brushmarks have to level out as much as possible and the polymer particles in the dry film must coalesce. Soluble polymers are added to adjust the viscosity and to control sedimentation. This is partly due to the increase in the medium viscosity as a result of the entanglements of the long polymer molecules but a major effect is for the polymers to induce a weak flocculation of the particles in a process known as *depletion flocculation*. Now, we must also understand how polymer molecules behave in solution, how they interact with particle surfaces and effect the particle–particle interaction forces.

The generic problems that we find when studying this coating are as follows:

(a) control of particle size (of both inorganic and organic polymeric particles);
(b) surfactant behaviour in solution and adsorption;
(c) drying and the redispersion of powders;
(d) solution properties of polymers;
(e) particle interaction forces and the effect of surfactants and polymers on these;
(f) sedimentation in concentrated systems;
(g) flow properties of concentrated systems.

2.2 Paper

Paper is another material of colloidal origin, which we use without a second thought. It may be in the form of newsprint, a cardboard box, a glossy magazine or the high-quality material that our degree certificates are printed on. It is formed from cellulose, a naturally occurring sugar-based polymer most frequently obtained from trees. When wood is pulped for the manufacture of paper, the cellulose is separated into fibres with sizes into the colloidal domain. The fibres are filtered to give a mat and dried in a high-speed continuous process. The fibres are negatively charged and this plays a role in the tendency of fibres to aggregate, with the latter being an important feature in the formation of a dense filter mat in which the particles are aligned to give maximum strength in the direction of the moving sheet. The understanding of both particle aggregation and filtration is paramount for successful production in high-speed modern equipment.

Pigments such as titanium dioxide are added to give a white sheet. As the fibres are hollow, some of the pigment particles end up inside the fibres. Removal of this can become a problem in recycling. Ink from printing on the exterior of the paper is less of a problem but does require the removal by detergent action of surfactant materials. The attachment and detachment of particles from surfaces require an understanding of the interparticle forces and how we can manipulate them, whether by chemical environment or surfactant type.

Glossy paper requires additional colloidal treatment. Well-dispersed kaolinite platelets are coated onto the surface and give a filler aligned parallel to the paper surface. Kaolinite has both negatively and positively charged surfaces, which tend to stick very firmly together to give a strong open particle network. This aggregation is controlled either by inorganic ions, such as phosphates, or organic polyelectrolytes and again the ability to manipulate interparticle forces is important. A binder is used with the clay surface to give a sealed, smooth and glossy final surface. A colloidal dispersion of polymer particles makes a suitable material. Emulsion polymerization is the normal route for this type of material. The application of the coating mix requires a knowledge of the flow of concentrated dispersions.

Some of the generic problems that we may identify here are as follows:

(a) control of particle–particle forces;
(b) separation of colloidal systems;
(c) interaction of surfactants with surfaces and detergent action in the removal of particulates;
(d) hetero-aggregation and its control;
(e) particle size control.

2.3 Electronic Inks

Modern hybrid circuits are built up from sequential printing of fine circuits and layers of insulating material. The circuits are printed by using inks with metallic colloidal particles dispersed in organic media. For example, gold or palladium has first to be produced as fine particles, separated and dried. Sufficient knowledge to enable the control of particle size and the subsequent separation of the colloidal particles is paramount here.

To make it into an ink suitable for printing, the system is dispersed in organic solvents with the aid of a surfactant to prevent the particles from sticking together. The mechanism of the stabilization must be understood. The viscosity of the concentrated dispersion has to be suitable for both flow during the screen-printing and the production of the correct film thickness. After drying, the circuits are completed by sintering the particles to give optimum conductivity. This process has parallel problems to film formation with polymer particles in other coatings, as well as in the firing of ceramic materials, whether these are derived from clays or other oxides such as those employed in high-grade ceramics used, for example, as chip bases in the electronics industry. The generic colloidal problems that we can immediately identify in this case are as follows:

(a) particle size control;
(b) separation and drying of particles;
(c) wetting of dry powders;
(d) adsorption of surfactants;
(e) stabilization of particles in a dispersion;
(f) control of flow properties;
(g) wetting of surfaces;
(h) sintering of fine particles;

2.4 Household Cleaners

A large amount of surfactant is sold for domestic cleaning purposes whether for clothes, skin or other surfaces. Each of these will have a different detailed formulation, of course, and as an example we will choose a cleaner for a surface such as a sink. The first requirement is that there is a high surfactant concentration. This is needed to solubilize grease and re-suspend particulate material. Hence, an understanding of detergent action is essential. Abrasive particles are required to break up the films that are responsible for staining but these particles should not be of such a size that they produce deep scratches or produce a 'gritty' feel. Particles of a micron or two in size will be satisfactory. The creamy feel is also achieved by the formation of long branching 'worm-like' assemblies of the surfactant molecules and requires a sufficient understanding of surfactant phase behaviour to optimize this.

The size and density of the abrasive particles are such that sedimentation will occur in a short period and to prevent this the system can be gelled by the addition of a soluble polymer. This has the side benefit of enhancing the texture or feel of the material. The solution behaviour of polymers and the control of the flow properties have to be understood in order to optimize the formulation. The generic problems here can be identified as follows:

(a) phase behaviour of surfactants in solution;
(b) detergent action;

(c) control of particle size;
(d) solution behaviour of polymers;
(e) control of flow properties.

2.5 Butter

Milk is a colloidal dispersion of fat droplets which are stabilized by the protein casein. This protein prevents the coalescence of the fat drops by a combination of electrostatic repulsion and a steric barrier as the protein layers make contact. On standing, the fat drops rise to the top in a process known as *creaming* which is analogous to sedimentation. So far, colloid stability and creaming (*sedimentation*) can be identified as areas of importance.

In the churning process, a phase inversion is produced and a water-in-oil emulsion is formed from an oil-in-water system. The saturated animal fats have a molecular weight such that they crystallize at temperatures close to body temperature. This is the reason why butter is difficult to spread at low temperatures. Many spreads are produced by blending in lower-molecular-weight vegetable oils with a lower melting point. The generic colloidal aspects are as follows:

(a) interaction forces between particles;
(b) coalescence of emulsion droplets;
(c) phase inversion of emulsions;
(d) flow behaviour of concentrated dispersions.

There are many other materials that are colloidal at some stage of their use but the colloidal problems can still be reduced to just a few generic problems. It is important to recognize this in spite of the complexity of a particular system. At first sight, it is often difficult to understand how the apparently abstract physics and chemistry presented in most courses and texts can apply to a 'practical system'. The application of the general principles though are usually sufficient to enable the problems to be both defined and tackled in a systematic manner. All of these points will be addressed in the following chapters.

3 CONCENTRATED COLLOIDAL DISPERSIONS

Traditionally, our ideas of colloidal interactions have stemmed from the behaviour of dilute systems of colloidal particles and the theoretical work based on two isolated particles interacting. This is nearly always in quite a different concentration region from the systems in which we employ colloids. However, in recent years this situation has changed and we now have a great body of work on concentrated dispersions. Of course, most of the academic work has been on model systems but general principles apply to the more complicated systems that are in everyday use.

As a starting point, it is important to describe what we mean by a dilute dispersion. This is not based on just the value of the weight or even the volume fraction. It is based on the mean separation of the particles compared to the range of the interaction forces between the particles. In the dilute state, the particles are well separated so that the particle interactions are negligible at the mean separation. The consequence of this is that the particles diffuse in a random fashion due to the Brownian motion, with a diffusion constant that can be described by Equation (1.4). The distribution of the particles in space can be considered as uniform, i.e. randomly distributed and the spatial correlations are very weak. Now, this is only strictly true for dispersions of particles which approximate to hard spheres. If there are either forces of attraction or repulsion acting between particles there will be some deviation from random as the particles collide. This point can be important but we do not need to consider it in detail at this stage; we only need to be aware of the possibility. In a fluid continuous phase, the motion of particles can be described by the hydrodynamics appropriate to an isolated particle. This is true for diffusion, sedimentation or viscous flow. The behaviour of the dispersion can be thought of as analogous to that of a gas except that the motion is Brownian and not ballistic, i.e. any two particles will experience many changes of direction before colliding. This means that the concept of a mean free path is difficult to apply.

If we now steadily replace the continuous phase by more particles, as the concentration increases our colloid becomes a condensed phase and we have a more complicated behaviour. This is a familiar concept to the physical scientist who will immediately recognize this behaviour as similar to that which occurs when a molecular gas is compressed until it forms a liquid and finally a solid. Many of the thermodynamic and statistical mechanical ideas translate well from molecular liquids to colloids in the condensed state. However, a little caution is required as the forces can be quite different. A liquid medium, for example, can result in hydrodynamic forces with a range of a few particle diameters. A very attractive feature though is that the colloidal forces can be readily manipulated by changes in the chemical environment of our colloidal particles. This, in turn, can dramatically alter the behaviour and thus it provides the means of manipulating the material to suit our needs more closely.

Now, in this condensed phase there will always be strong interactions between the particles. This is the case whether the interactions are repulsive or attractive. Such a situation gives rise to strong spatial correlations and we have a shell of nearest neighbours. The number of particles in this shell is the coordination number and this reflects both the magnitude and type of force as well as the concentration or particle number density. For example, if the particles are of very similar size and the forces are repulsive, colloidal crystals can be formed with very long-range order. The spatial arrangement is face-centred cubic and if the lattice spacing is of the order of the wavelength of light, strong diffraction will be seen. Opal is a naturally occurring colloid where this effect is utilized as a gemstone. When the particles are in a liquid medium, 'exciting behaviour' can be seen. Three modes of diffusive motion can be identified. The particles are all moving due to the thermal or Brownian motion but are generally constrained to be within their individual coordination shell. This motion is quite rapid and is known as *short-time self-diffusive motion*. The motion is still random and, if we were to take a series of 'snapshots' of a particular volume, we would see that the number density of particles in that region would fluctuate about the global mean for the dispersion. The diffusion of these regions is the *collective diffusion* and the constant is slower than for short-time self-diffusion. All liquids behave in this way and it is this local density fluctuations in the continuous phase that produces the Brownian motion of the particles. Occasionally, the fluctuations will allow sufficient separation in a coordination shell for a particle to move through and change its neighbours. This is known as *long-time self-diffusion*.

The flow properties reflect this interesting behaviour. To illustrate the point, let us consider a simple system of uniform particles with strong repulsive forces at a high concentration. The particles are highly spatially correlated in a face-centred cubic structure. If we deform the structure, the arrangement of particles is distorted. We have had to do work on the structure and the energy is stored by the movement of the particles to a higher-energy configuration. An elastic response is observed. Over time, the particles can attain a new low-energy configuration in the new shape by the long-time self-diffusion mechanism. The system now will remain in the new shape without applying the external force, i.e. the structure has relaxed and the elastically stored energy has dissipated (as heat). This is known as the *stress relaxation time* and the material is behaving as a *viscoelastic* material. In other words, we are saying that the material is now exhibiting a 'memory' and it takes several relaxation times before the original shape is 'forgotten'. When this timescale falls within that of our normal perception we are aware of the textural changes and many concentrated colloids are manipulated to take advantage of this.

The transition from a dilute to a condensed phase can be very sharp and is a function of the range of the forces, as noted above. We may now move back to consider a system of hard spheres – a system, incidentally, which can only really be attained in a computer simulation but which we can get quite close to under very limited conditions. In a computer simulation it is possible to take a fixed volume and increase the fraction of that volume which is occupied by particles, all in random Brownian motion, of course. The volume fraction of the 'dispersion' is simply the product of the number of particles per unit volume, N_p, and the particle volume, v_p, as follows:

$$\varphi = N_p v_p \tag{1.6}$$

The simulations show that a liquid/solid transition occurs at $\varphi_t \sim 0.5$. Below this transition we have a viscoelastic liquid and above it a viscoelastic solid. How does this relate to systems with colloidal particles stabilized by long-range electrostatic repulsion or extensive polymer layers preventing the particles from coming together? We can introduce the concept of an *effective volume fraction* which is calculated from the particle volume which has been increased by a volume from which neighbouring particles are excluded due to repulsion. For example, we can easily visualize the case for a dispersion of spherical particles, each of which has an attached polymer layer which physically prevents approach of another particle. Figure 1.1 illustrates this schematically.

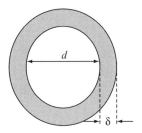

Figure 1.1. Schematic of a particle with an adsorbed polymer layer which increases the effective volume fraction of the system.

The thickness of the polymer layer is denoted by δ which gives the effective hard sphere diameter as $(d + 2\delta)$. The effective hard sphere volume fraction is now:

$$\varphi_{HS} = N_p \frac{\pi(d + 2\delta)^3}{6} \tag{1.7}$$

and the liquid/solid transition would fall to a lower value of the volume fraction calculated from the core particles. Thus:

$$\varphi_{HS} \sim 0.5$$

so:

$$\varphi_t \sim 0.5/(\varphi_{HS}/\varphi)$$

and then:

$$\varphi_t \sim \frac{0.5}{\left(1 + \frac{2\delta}{d}\right)^3} \tag{1.8}$$

When the stability is due to long-range electrostatic repulsion between particles, we may also define an effective hard sphere diameter. The simplest approach in this case is to recognize that the principle of the equipartition of energy applies to colloidal particles so that a particle moves with $k_B T/2$

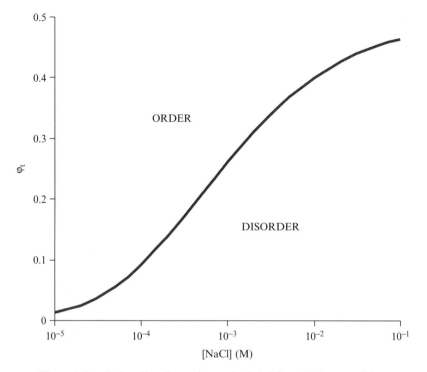

Figure 1.2. Order–disorder regions calculated for a 100 nm particle.

kinetic energy along each of the x, y and z coordinates. Thus, an *average* value of the energy of a Brownian collision would now be $k_B T$. We may then take the distance δ as the distance at which the repulsive energy reaches this value and again define an effective hard sphere diameter as $(d + 2\delta)$. This now enables us to try to estimate the concentration of the liquid/solid transition. Figure 1.2 illustrates the result for a particle with a radius of 100 nm. We will return to this in more detail in a later chapter but we should note at this point that because the electrostatic interactions are relatively 'soft' the material will form a *soft* solid. That is, the application of an external force can cause large deformations. This is a natural consequence of the range of the interparticle interactions compared with the particle size. The farther we move to the right in Figure 1.2, then the harder the solid becomes.

4 INTERFACES

As soon as we consider a fine dispersion of one phase in another the issue of the interface between the two phases becomes of major importance. As an illustration of the points that arise, consider the atomization of water into fine droplets in air. The area per unit mass is known as the *specific surface area* (SSA). The disperse phase is in the form of spherical particles because there are surface tension forces that we will discuss in a moment. The calculation of the SSA is based on the area of a sphere of diameter d (πd^2) divided by its mass $((\pi d^3/6)\rho_{H_2O})$, where ρ_{H_2O} is the density of water. This gives:

$$SSA = \frac{6}{d\rho_{H_2O}} \qquad (1.9)$$

Thus, for a litre of water (i.e. about 1 kg) before atomization, the SSA $\sim 0.05\,m^2$. After spraying to give droplets of 1 μm, the value of the SSA is $\sim 6 \times 10^3\,m^2\,kg^{-1}$ and we are now dealing with an interfacial area larger than the area of a football field! It is easy to see why the effectiveness of a catalyst is maximized when in a finely divided form, and also why the oxidation of finely divided materials such as metals can be a dangerous problem due to the exothermic reaction becoming uncontrollable. If the droplet size were reduced to the order of 10 nm, the specific surface area would be $\sim 10^6\,m^2\,kg^{-1}$. It is interesting now to consider the fraction of the molecules that would be at the interface as the size of the drop is made smaller. The approximate values are shown in Figure 1.3 and are significant fractions for drops in the colloidal size range – particularly when the droplets would be in the nanoparticle size range, i.e. up to a few tens of nanometres in diameter. This looks just like a simple exercise in geometry so far but the implications are quite important. To illustrate this, let us think about the amount of work we would have to do to take our 1 kg of water down to droplets of 300 nm in diameter where $\sim 0.1\%$ of the water molecules are at the surface. Remember that the intermolecular forces in water are dominated by hydrogen bonding – giving the tetrahedral structure – and at 4 °C when the density is 1000 kg m^{-3} this would be nearly complete. Thus, if we make the crude assumption that each surface molecule is one hydrogen bond short and that the energy of a hydrogen bond is $\sim 40\,kJ\,mol^{-1}$, then we may estimate how much work we would have to do to disperse the water into a fog. (Note that there is a factor of 2 involved as each hydrogen bond broken would result in two fresh surface areas.) This result is also illustrated in Figure 1.3. Of course, if we had broken all of the hydrogen bonds, we would have boiled the water (this would take $\sim 2.5 \times 10^3\,kJ$) but a lot of work is required to get bulk water down to drops in the sub-micron region.

The above illustrates that we have to do work to create a new surface and that the origin is the work done against the intermolecular forces. This is a key concept when we consider surfaces or interfaces. Here, the term 'surface' is taken to refer to a surface of a liquid or solid in contact with a gas or vapour, while the term 'interface' is used to describe the region between two

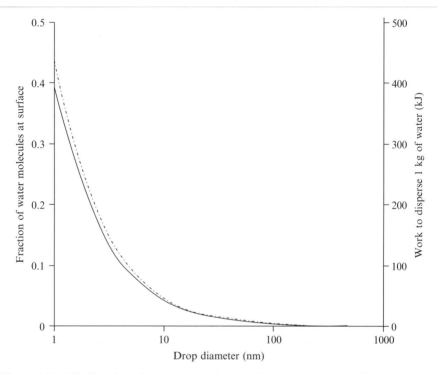

Figure 1.3. The fraction of water molecules in a drop that are located at its surface: (——) fraction of water molecules at surface; (—·—·—) work to disperse 1 kg of water.

condensed phases whether two liquids, two solids or a liquid and a solid. In the bulk of the condensed phase, the intermolecular forces act between the atoms or molecules in essentially a symmetric fashion. At the surface or interface, there is an imbalance as the local chemical environment changes. If we think of the intermolecular forces as molecular springs, the imbalance in attractive force results in a surface tension, γ_1. This acts to minimize the surface area. Now, when the surface area of the liquid is increased by an amount ∂A against this surface 'spring' tension, the amount of work is given by the following:

$$\partial W = \gamma_1 \partial A \qquad (1.10)$$

This is only the case for a pure material. If there are dissolved species present, we must consider the presence of such species at the surface or interface as we shall see when we explore surfactants. The units of the surface tension are $J\ m^{-2}$ (i.e. energy per unit area) and as energy is force multiplied by the distance moved, the dimensions are also written as $N\ m^{-1}$, which is the spring constant. Water, for example, has a value for γ_l of 72 mN m^{-1}. If we integrate Equation (1.10) up to an area of 1 m^2, we have the energy required to create a fresh surface of unit area, and we see that if this area is the SSA of droplets of 300 nm in diameter, we require 1.4 kJ. This value compares favourably with the simplified estimate illustrated in Figure 1.3.

In water, the hydrogen bonding interaction is the strongest intermolecular force although it is not the only contribution. The usual van der Waals forces also play a role and contribute about 25 % of the surface energy of water. These are the forces that cause an interaction between all atoms and molecules, even the inert gases. They are the *London dispersion forces* which are due to the coupling of the fluctuations of the charge density of the electron clouds of one atom with its neighbours. This will be discussed in some detail in Chapter 3 with aspects of the surface energy being discussed in Chapter 6. An important feature of the recognition that an appreciable amount of work is required to generate new surfaces is that the process is endothermic and that the dispersed state is not the lowest energy condition. In other words, there is a natural tendency for droplets to coalesce and for particles to aggregate. To maintain the material in the colloidal state, we must set up the correct conditions.

We have just begun to explore the molecular implications of an interface or surface. The structure of the liquid surface in equilibrium with its vapour cannot be as well defined as that of a crystalline solid and the concept of a well-defined plane is a convenience rather than a reality as there is really an interfacial region. When a surface is expanded or contracted, diffusional motion to or from the bulk accompanies the changes and the intensive properties of the interface remain unchanged. With a solid surface, the situation can be more complex and crystal structure, for example, can result in anisotropy. The surface free energy described above appears to be straightforward. However, equating the surface free energy just with the surface tension can only hold for a pure liquid. Whenever another species is present, the distribution becomes important as this controls the details of the intermolecular forces in the interfacial region. If the concentration of solute species is lower in the surface region than in the bulk phase, the species is termed *lyophilic* as it 'prefers' the bulk phase to the surface phase. The solute species is *negatively adsorbed* at the surface or interface. Indeed, the stronger interaction between the lyophilic solute species and the solvent can even lead to a small increase in the surface tension. If the molecules tend to accumulate at the interface they are termed *lyophobic*. This tendency for the solute species to accumulate at the interface implies that the intermolecular interactions are most favourable if there is a separation of the solvent and solute into the region of the surface. This is particularly marked for *amphiphilic* (also termed *amphipathic*) molecules. These are a class of molecules known as *surfactants* or *surface active agents*. In this case, there are two distinct moieties making up the molecule:

part of the molecule is lyophilic while another part is lyophobic. In water, a polar group such as the salt of a carboxylic acid group would be a lyophilic moiety. In water, this is also described as being *hydrophilic*. A linear paraffin chain or an aromatic hydrocarbon would be a typical lyophobic, or *hydrophobic*, moiety. The increase in concentration at the interface is known as the *surface excess*.

The surface tension of water is lowered as solute molecules accumulate in the surface region. Water is an associated liquid and the solute molecules do not display the relatively strong hydrogen bonding forces. Thus, even if the London dispersion forces are stronger, the surface tension is lowered. A diagramatic picture of the surface of a solution is shown in Figure 1.4. Of course, this picture is not restricted to the surface of an aqueous solution.

There are some important ideas illustrated in this figure. The interface between the liquid phase and the vapour phase is not a plane when we work at the molecular level. Rather, it is a region a few molecules in thickness – say five or six – where the molecular density or concentration profile changes from that of the liquid to that of the vapour. Hence, we can think of there being a surface phase. When there are two molecular species present, we can expect the concentrations to vary with the nature of the solute species, as indicated in the previous paragraph. In this figure, we have large solute molecules which are lyophobic and so there is a surface excess concentration. This is illustrated by the peak in the concentration profile (Figure 1.4(a)), and as shown the large molecules have a much lower vapour pressure than the solvent molecules, but this, of course, is not a prerequisite. When we know the local concentration, in principle we can estimate the surface tension. Direct measurement of the concentration profiles is not something that has been achieved with precision so far but it is possible to estimate the surface excess from measurements of the surface tension. To do this, we need to use just a little thermodynamics, as clearly laid out in the text by Everett [1].

First, we are going to choose a volume for our system at equilibrium which contains saturated vapour, v, the solution phase, ℓ, and the surface phase, s. Our problem is to define the volume of this surface phase. What we are going to do is to model it as though it were just a planar surface with all of the material in the surface phase located in that plane. This plane is known as the *Gibbs dividing surface* – the Gds line in Figure 1.4(a) – and for simplicity we will consider a volume with unit area Gds, as in Figure 1.4(b). As this is a model, we may choose the location of the Gds to be the most convenient, i.e. to make the calculations as simple as possible. The appropriate concentration terms are defined as follows:

Γ_{1s} is the number of moles of solvent species per unit area at the Gds;
Γ_{2s} is the number of moles of solute species per unit area at the Gds;
$c_{1\ell}$ is the concentration of solvent in the liquid phase;

(a)

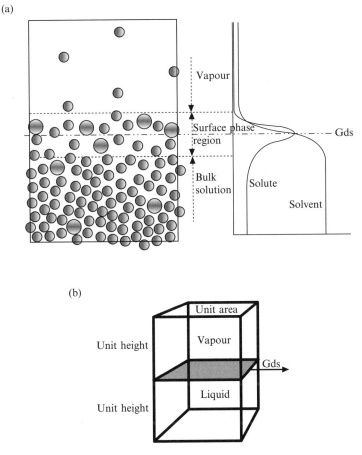

(b)

Figure 1.4. Representations of a simple model for the liquid–vapour interface; Gds indicates the Gibbs dividing surface (see text for details).

c_{1v} is the concentration of solvent in the vapour phase;
$c_{2\ell}$ is the concentration of solute in the liquid phase;
c_{2v} is the concentration of solute in the vapour phase;
c_1 and c_2 are the total concentrations of solvent and solute in the system, respectively.

Thus, we have:

$$c_1 = c_{1\ell} + c_{1v} + \Gamma_{1s}, \text{ and } c_2 = c_{2\ell} + c_{2v} + \Gamma_{2s}$$

i.e.

$$\Gamma_{1s} - c_1 = -(c_{1\ell} + c_{1v}), \text{ and } \Gamma_{2s} - c_2 = -(c_{2\ell} + c_{2v})$$

$$\frac{\Gamma_{1s} - c_1}{(c_{1\ell} + c_{1v})} = \frac{\Gamma_{2s} - c_2}{(c_{2\ell} + c_{2v})}$$

$$\Gamma_{2s} = c_2 + (\Gamma_{1s} - c_1)\left(\frac{c_{2\ell} + c_{2v}}{c_{1\ell} + c_{1v}}\right)$$

which gives:

$$\Gamma_{2s} - \Gamma_{1s}\left(\frac{c_{2\ell} + c_{2v}}{c_{1\ell} + c_{1v}}\right) = c_2 - c_1\left(\frac{c_{2\ell} + c_{2v}}{c_{1\ell} + c_{1v}}\right) \tag{1.11}$$

In principle, the latter term is experimentally accessible but we can simplify Equation (1.11) if we choose the location of our Gibbs dividing surface carefully. We will define this surface so that the excess number of solvent molecules on the vapour side is exactly matched by the deficit on the liquid side. This gives the value of Γ_{1s} as 0 and then we call the surface excess of the solute, Γ_{2s}, *the relative adsorption of solute* at the surface.

The Helmholtz free energy of the system is just the sum of the free energy of each phase:

$$F = F_v + F_\ell + F_s \tag{1.12}$$

The surface term is of importance for our colloidal systems where the surface area is large. For the bulk phases, we have the usual equation for the change in free energy with the amount n of species i:

$$dF_v = -S_v dT_v - p_v dV_v + \sum_i \mu_{vi} dn_{vi} \tag{1.13}$$

and an analogous equation for the surface:

$$dF_s = -S_s dT_s + \gamma_s dA_s + \sum_i \mu_{si} dn_{si} \tag{1.14}$$

Here, the pressure term is now the surface tension and the sign has to change as it is a tension instead of a pressure. The phase volume is replaced by the area of the surface. The temperature is constant and so when we integrate equation (1.14) we then obtain the Gibbs–Duhem Equation for the surface:

$$F_s = \gamma_s A_s + \sum_i \mu_{si} n_{si} \tag{1.15}$$

Differentiating Equation (1.15) generally gives us:

$$dF_s = \gamma_s dA_s + A_s d\gamma_s + \sum_i \mu_{si} dn_{si} + \sum_i n_{si} d\mu_{si} \qquad (1.16)$$

We can now equate Equations (1.16) and (1.14), recalling that the 'SdT' term is zero as we are working at constant temperature, to give the following:

$$A_s d\gamma_s + \sum_i n_{si} d\mu_{si} = 0 \qquad (1.17)$$

Dividing through by A_s gives us the relative adsorption of the components as follows:

$$d\gamma_s = -\sum_i \Gamma_{si} d\mu_{si} \qquad (1.18)$$

With a system with just two components, we can choose the Gds to give $\Gamma_{s1} = 0$ and so remove the solvent from the equations. In addition, it is convenient to use the chemical potential of the solute in the liquid phase (at equilibrium, the chemical potential of each species, μ_i, is the same in each phase) and we have the *Gibbs adsorption isotherm*, as follows:

$$d\gamma_s = -\Gamma_{s2} d\mu_{\ell 2} \qquad (1.19)$$

The chemical potential is related to how much of the solute we have in the liquid phase, that is, the activity of the solute:

$$d\mu_{\ell 2} = RT \ln a_{\ell 2} \qquad (1.20)$$

This now gives us a convenient means of estimating the relative adsorption of the solute at the surface by measuring the slope of the curve of the surface tension as a function of the natural logarithm of the activity:

$$\Gamma_{s2} = -\frac{1}{RT} \left(\frac{d\gamma_s}{d \ln a_{\ell 2}} \right) \qquad (1.21)$$

This equation is frequently used to estimate the amount of strongly adsorbed material such as surfactants at the liquid surface. It will only be approximate if the molar concentration is used as even though the solution concentrations are usually low there are problems such as these are far from being ideal solutions with an activity coefficient of unity. When there are several components present, the algebra is only a little more complicated and general expressions can again be found in the text by Everett [1].

5 SURFACTANTS

Surfactants are molecules which have a chemical structure which makes it particularly favourable for them to reside at interfaces. Hence, they are termed *surface active agents*, or simply *surfactants*. Such molecules are a frequent component of colloidal systems, whether man-made or naturally occurring, and so it is of great importance to know how much resides at the interfaces in our systems. It was shown above that the rate of change of surface tension with the logarithm of the activity gives us an estimate of the amount of the solute adsorbed at the interface. Now, we should use Equation (1.21) to make all of the above algebraic manipulation worthwhile and to get a feel for what the equation can tell us. The example that we will use is the experimental data plotted in Figure 1.5 for a simple cationic surfactant in water. The surfactant in this case is hexadecyltrimethylammonium bromide ($C_{16}TAB$). This consists of a straight 16-carbon aliphatic chain with the quartenary ammonium group as the terminal group at one end. The ionic terminal group carries a positive charge and is strongly solvated so that the long aliphatic chain is carried into solution in water. The solution behaviour of such surfactant molecules will be discussed in more detail in Chapter 2, but represents a good example for our current purpose. An aliphatic chain of 16 carbon atoms is not very soluble in water and the result is that there is strong adsorption at the water–vapour interface. The experimental curve of surface tension against the concentration is given in Figure 1.5. The surface tension shows a monotonic decrease up to a concentration of $9 \times 10^{-3}\,\mathrm{mol\,l^{-1}}$.

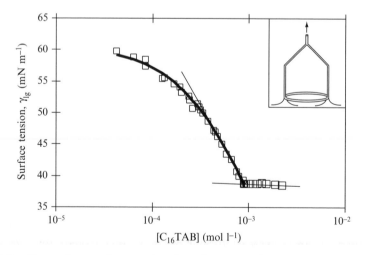

Figure 1.5.   The surface tension as a function of concentration for heaxadecyltrimethy-lammonium bromide in water.

Beyond this, the curve is almost parallel to the x-axis. This point at which this abrupt change in slope occurs is known as the *critical micelle concentration* (cmc). We will come to this shortly but let us concentrate on the first section of the curve. First, we must recognize that we are using molar concentrations and *not* activities. Although the concentrations are low, the activity coefficient will be slightly less than 1. Thus, our results will only be approximate but still of use. The limiting slope of the curve prior to the cmc is 1.18×10^{-3}, which yields a value from Equation (1.21) for Γ_{s2} of $4.6 \times 10^{-6}\,mol\,m^{-2}$. At 35 °C, we have the area occupied by a $C_{16}TAB$ molecule as $0.36\,nm^2$. This is about twice that found for an undissociated fatty acid which gives a close-packed layer at $0.19\,nm^2$. The first thing to note is that the trimethylammonium head group is a larger group than a carboxylic acid group, but twice as big? Well, perhaps not. So, the second feature that we should consider is that the group is positively charged. Like charges repel and this acts to reduce the packing density.

Let us now consider the charge in more detail. We have a surface for which we estimate from the surface tension measurements that there would be a positive charge (i.e. $1.6 \times 10^{-19}\,C$) for every $0.36\,nm^2$ of surface. This gives a measure of the *surface charge density*, σ_s, of $\sim 45\,\mu C\,cm^{-2}$. Experiments with solids, such as silver iodide, or oxides, such as titanium dioxide, yield surface charge densities in the range $1 - 15\,\mu C\,cm^{-2}$, and so this clearly would be a very highly charged surface. Of course, the head groups are just one half of the ion pair, while the bulky bromide ion is the counter-ion to the surface charge and will be strongly attracted to the positively charged surface. The binding of the counter-ions reduces the repulsion between the head groups. The charge on the surface attracts the counter-ions but, as the concentration of the latter is high, diffusion acts in the opposite direction, tending to dilute the concentration at the surface. The model for the surface now consists of the hexadecyltrimethylammonium ions located in the surface with the hydrocarbon tails extended into the vapour phase and the head groups in a densely organized layer which is highly charged. The charge is balanced by many counter-ions which are closely bound to the surface with the remaining counter-ions in a more diffuse layer where the remaining electrostatic attraction is balanced by diffusion. This concept of a charged surface with a layer of counter-ions, some of which may be strongly bound, and the remainder in a diffuse array is a key concept which helps us to understand the behaviour of charged particles in a dispersion. This is known as the *electrical double layer* and will be discussed more fully in subsequent chapters.

This is an appropriate point at which to discuss the measurement of the tension of the surface. The data presented in Figure 1.5 were obtained by measuring the force exerted when attempting to pull a platinum ring out of the surface. The equipment used for this was a DuNoüy tensiometer, although this is just one approach. Chapter 6 gives details for several other methods. The inset shown in Figure 1.5 illustrates the geometry of the measuring element. As a force is exerted on the ring support perpendicular to the surface, the surface resists the displacement of the ring. In principle, the force at which the ring will detach is given by the surface tension in $N\,m^{-1}$ multiplied by twice the circumference of the ring (in m). (Remember that the surface makes contact with both sides of the platinum wire of the ring). A computer-controlled microbalance does this job very well. However, the points that we need to keep in mind here arise from the usual condition in thermodynamic calculations that at some point we have required the system to be at equilibrium. Thermostatting is of course a prerequisite. The first problem that we must take care with is that the vapour phase should be saturated. Hence, our system should be enclosed and sufficient time taken for the vapour phase to come to equilibrium. This is particularly important if the vapour pressure of the solute is significant when compared to the solvent. This is not a problem with large molecules such as $C_{16}TAB$ though. The second problem of equilibrium is, however, that at low concentrations of surfactant a significant time passes before the molecules in solution diffuse to the surface and equilibrium becomes established. Each point of the curve shown in Figure 1.5 usually follows a dilution of the solution and mixing. At concentrations close to the cmc, there are many surfactant molecules close to the surface and equilibrium is quickly attained. However, at the other end of the curve several minutes are needed for consistent measurements to be achieved, repeat readings are necessary to confirm the values and the time taken to produce the full curve can stretch into hours!

The slope of the surface tension–log (concentration) curve increases steadily as the surfactant concentration is increased. This tells us that the relative adsorption of the $C_{16}TAB$ is increasing as more is added to the water. However, at the cmc there is an abrupt change in slope and what occurs now is that the surface tension changes very little with more concentrated solutions. What we find here is that above the cmc, where the surface is closely packed, there are small aggregates of surfactant molecules in solution. In other words, surfactant in excess of that required to give a concentration equal to the cmc has self-assembled into 'macro-ions'. Typically, the aggregation number of surfactant molecules in a micelle is around 50–100 close to the cmc, with diameters of a few nanometres. The core of the micelle can be pictured as rather like a small oil droplet with the surfactant head groups located at the surface. The latter moieties are strongly hydrated and the first two or three carbon atoms of the tail near to the head group are close enough to be influenced by the head group hydration. In fact, on the nanometre scale the concept of a clear distinction between the outer edge of the hydrocarbon core and the aqueous phase breaks down. This ability for surface active species to

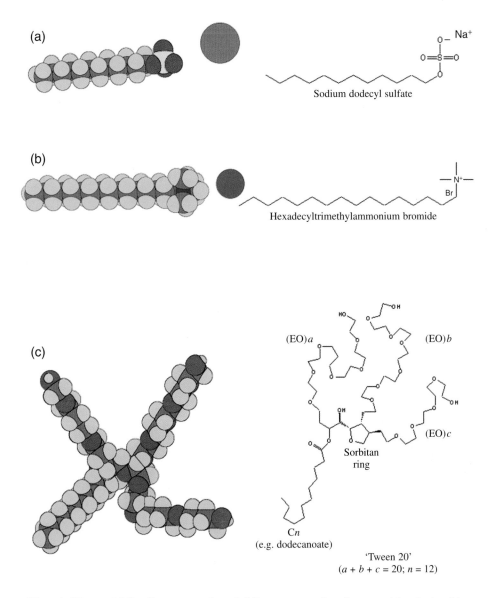

(a)

Sodium dodecyl sulfate

(b)

Hexadecyltrimethylammonium bromide

(c)

(EO)a

(EO)b

(EO)c

Sorbitan
ring

Cn
(e.g. dodecanoate)

'Tween 20'
($a + b + c = 20$; $n = 12$)

Plate 1 (Figure 2.14). Some examples of different types of surfactants: (a) anionic; (b) cationic; (c) non-ionic (EO, ethylene oxide).

self-assemble into various structures is extremely important in a wide range of applications, from cell membranes to washing clothes.

It is also possible to use the variation in surface tension with surfactant concentration to monitor the adsorption of the surfactant onto the surfaces of particles in suspension. At equilibrium concentrations up to the cmc, the procedure can be similar to a titration where a surfactant solution of known concentration is added and the surface tension monitored without separating the solids from the liquid. However, beyond the cmc the phases must be separated, for example, by centrifugation, and an aliquot of the supernatant removed and diluted carefully to below the cmc prior to the measurement. The data presented in Figure 1.6 show the adsorption isotherm of $C_{16}TAB$ onto a sample of china clay. For comparison, data obtained from radiochemical assay are also given. The faces of the clay particles were negatively charged and the edges positively charged at the pH of the experiment and so the adsorption occurs on the particle faces. The isotherm shape is typical of that of an high-affinity isotherm. Initially, the attachment is by the head groups of the surfactant molecules leading to a monolayer, which results in an hydrophobic surface and further adsorption occurs to give a bilayer. This coverage occurs at an equilibrium concentration of the surfactant in the solution which is approximately half the value of the cmc. At much higher concentrations, there is evidence of yet further adsorption. The clay surfaces are not simple though as they possess 'steps' and the adsorption close to the step edges may require higher equilibrium concentrations. However, prior to

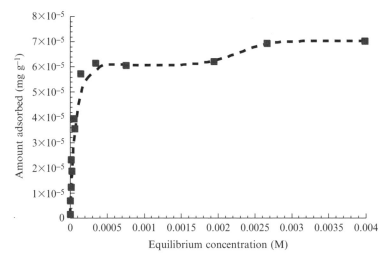

Figure 1.6. The adsorption isotherm of heaxadecyltrimethylammonium bromide on sodium kaolinite at $35\,^{\circ}C$; data for adsorbed amounts below $4 \times 10^{-5}\,mg\,g^{-1}$ were obtained by independent radiochemical assay measurements.

the adsorption of the surfactant, the clay particles are aggregated edge-to-face in a 'house of cards' structure. As soon as the adsorption plateau is reached, the particles are completely dispersed and the surfactant titration technique is well suited to providing this type of adsorption data rapidly. At the plateau, the area occupied by each molecule (calculating the face area from the specific surface area measured by gas adsorption and reducing this by the fraction corresponding to the edge area) is $\sim 0.5\,nm^2$ in each layer. (Note that this is quite close to that found at the air/water interface at the same equilibrium concentration.)

One of the main uses of surfactants is to provide stability to dispersions of colloidal particles and the above titration technique provides a quick method to determine how much surfactant is required. However, the molecules are only physisorbed and *not* chemisorbed, and so care has to be taken when additions to the system are made. If the system is diluted with solvent, then surfactant will desorb until a new equilibrium is attained. To prevent this, dilution should be carried out with a solvent phase containing the equilibrium concentration of surfactant required to maintain the value where the adsorption plateau occurs. In addition to the provision of colloidal stability, surfactants are also used to aid the wetting and hence the dispersion of powders in liquids, as well as aiding the break-up of oil droplets in emulsification processes, as we shall see in later chapters.

6 SOLUTION POLYMERS

Macromolecules or polymers, like surfactants, are often a key component in colloidal systems and so it is important to introduce them here in this early part of the text. The robustness of the stability against aggregation of many colloids of biological origin is due to the presence of proteinaceous macromolecules on their surfaces. As an example of this we have to look no further than the stabilization of the fatty acid droplets in milk which are stabilized by casein. We often add polymers which will adsorb onto particles for this purpose. However, nature has provided a very effective mechanism for keeping particles apart by three components. Only part of the macromolecule adsorbs, i.e. is attached. This leaves the rest which is solvated to expand away from the interface and prevent other particles from close approach. The proteins are also charged and the charges repel other particles too, thus adding to the effectiveness of the stabilizing layer.

Synthetic polymers are also used as stabilizers. Homopolymers are not much use as stabilizers, as if they are readily soluble in the continuous phase they will not form strong effective attachments to the surface. Hence, we emulate the smaller molecules like surfactants and make the polymers contain some lyophobic blocks along the chain. Frequently, these polymers are of relatively low molecular weight, typically in the range of 5×10^3 to 50×10^3.

Polymers of higher molecular weights are also in common use though. These are employed to alter the flow or sedimentation behaviour of colloidal systems. For this reason they are termed 'thickeners' or 'rheology modifiers'. A polymer in solution increases the viscosity of that solution and high-molecular-weight material is particularly effective at this so that only a small amount is required. When molecular weights $> 10^6$ are utilized, however, problems in rheological behaviour become apparent. For example, droplets do not break away from the bulk cleanly – we have a 'stringy' behaviour which is due to a marked resistance to stretching. That is, the extensional viscosity is high and applications such as spraying become difficult. One solution to this problem is to use a lower-molecular-weight polymer and make it behave like a system of much higher molecular weight under quiescent conditions, but like a lower-molecular-weight material upon application. This is achieved by having a small mole percentage of lyophobic polymer material on the backbone of the polymer, which results in a weak assembly of these regions so that all of the polymer molecules are associated with each other. This has similarities to the self-assembly of surfactant molecules and is becoming increasingly widely utilized.

It is interesting to note that when soluble polymers are added as a rheology modifier to a colloidal dispersion, *a synergistic* effect is often observed. That is, the relative increase in viscosity of the dispersion is markedly greater than the relative increase found for the polymer solution on its own. What occurs here is that solution polymer, which does not adsorb to the disperse phase, produces a weak reversible aggregation of the disperse phase and this increased interaction is observed as a further change in the rheological behaviour.

Polymers with charged groups are known as *polyelectrolytes* and these can be added as stabilizing agents for particulate dispersions or to cause aggregation. For example, poly(acrylic acid) produces a good dispersion of china clay by adsorbing onto the edges which carry a positive charge. Positively charged polyacrylamide can be used to remove negatively charged particulates by a bridging mechanism which holds particles together and makes them easy to separate. The polymer concentration required to do this is extremely low. Too high a level could give complete coverage of the surfaces by the cationic polymer and provide (unwanted) stability of the system.

7 SUMMARY

This introduction has defined what we mean by colloidal systems and has illustrated how widely different systems can fit into this form of matter. The related systems of surface active molecules and macromolecules have also been introduced and shown how they are intimate adjuncts to colloidal dispersions. A few common systems have been described which, although they appear to be widely disparate, have some basic or generic aspects. These will be a focus of this text and will show why the subject has a marked interdisciplinary flavour.

REFERENCES

1. D. H. Everett, *Basic Principles of Colloid Science*, The Royal Society of Chemistry, London, 1988.

Chapter 2

Macromolecules and Surfactants

1 INTRODUCTION

Macromolecules and surfactants are added to nearly all formulations that include colloidal particles. Macromolecules may be added to control the rheological behaviour as well as to control the stability of the dispersion. Surfactants can be used as wetting agents to disperse dry powders, emulsifiers for oil-containing formulations, and to provide stability against aggregation, as well as being added as cleaning agents. Indeed, these molecules are so important to colloidal systems that we need to discuss their solution properties prior to the discussion of colloidal particles.

2 MACROMOLECULAR DEFINITIONS

The term 'macromolecules' is used here to include synthetic polymers such as poly(ethylene oxide), naturally occurring macromolecules, such as proteins like gelatin, or polysaccharides like ethylhydroxy cellulose, or oligomers such as cyclodextrin. In each case, the monomer or building block of the macromolecule is a small molecule. With synthetic polymers, the chains are often hydrocarbons with side groups which give the correct properties. For example, poly(acrylic acid) is water-soluble because of the polar carboxylic acid group on every second carbon atom along the chain, whereas polystyrene is soluble in aromatic hydrocarbons and not water as there is a benzene ring attached to every second carbon atom. Sugar rings – glucose is a common example – are the monomeric units of the polysaccharides, while amino acids are the building blocks of proteins. We use the term 'oligomer' to indicate that there is only a small number of monomeric units that are linked – may

Colloids and Interfaces with Surfactants and Polymers – An Introduction   J. W. Goodwin
© 2004 John Wiley & Sons, Ltd   ISBN: 0-470-84142-7 (HB)   ISBN: 0-470-84143-5 (PB)

be ten or twenty, for example. Some chains are simple linear molecules while others may be branched. This branching is taken to an extreme with *dendrimers* where each chain branch branches again and again to give a large approximately spherical unit. Chains can be cross-linked, and if this is achieved by covalent bonds, swollen gel particles can be prepared and these are termed *microgels*. If the chains are lightly cross-linked by placing a small amount of insoluble species on the chains, we have a weak self-assembly and have synthesized an 'associative thickener'. With a higher level of self-association, we can produce highly swellable gels such as the 'super absorbers'.

If there are N segments bonded together in one chain (i.e. the degree of polymerization is N) and the molar mass of each segment (monomer) is M_m, the molecular weight of that chain is given by the following:

$$M = NM_m \qquad (2.1)$$

During the polymerization process, a distribution of chain lengths is always produced [1]. Usually, the distribution is broad but some ionic-initiated polymerizations can be controlled to give a narrower distribution than, for example, a free-radical-initiated polymerization of a vinyl monomer. Hence, we need to define the the various kinds of average molecular weight:

Measured by:

Number average: $M_n = \sum_i \dfrac{M_i n_i}{n_i}$ Osmotic pressure

Viscosity average: M_v Viscometry

Weight average: $M_w = \sum_i \dfrac{M_i w_i}{w_i} = \sum_i \dfrac{M_i^2 n_i}{M_i n_i}$ Light scattering

Z-average: $M_z = \sum_i \dfrac{M_i^3 n_i}{M_i^2 n_i}$ Ultracentrifugation   (2.2)

These averages increase in the order $M_n < M_v < M_w < M_z$, and so it is important to define which method has been used to determine the molecular weight. Although the detailed distribution is often not known in detail, the width of the distribution is often characterized by the poly-dispersity, P, which is defined in terms of two of the commonly measured averages:

$$P = \frac{M_w}{M_n} \qquad (2.3)$$

Polymer molecular weight standards, used for calibrating equipment, for example, would have a value of $P < 1.1$, but polymers for bulk usage usually have a polydispersity with a value of 3 or more. It should be noted though that even a value of $P = 1.1$ represents quite a broad distribution and will, of course, also depend on the details of the 'skew' in that distribution. In practical usage of polymers, for example, as thickeners, this wide distribution can be useful as the changes in viscosity with rate of shearing the system is slower if the distribution is broad. Any property which is dependent on the diffusive motion of the components will be affected similarly.

3 CONFORMATION IN DILUTE SOLUTIONS

The texts by Flory [2, 3] present the classical descriptions of the solution properties of polymers in dilute solution, while other important texts include those by Yamakawa [4], deGennes [5] and Doi and Edwards [6]. The starting point for the description of the conformation of a large polymer molecule in solution is to use the statistics of a three-dimensional random walk. At this stage, the problem is simpler than a description of random motion, such as diffusion, because the step sizes are equal as each step has a dimension equal to the monomer unit in the chain. By considering each bond as a vector and summing the squares, the mean-square distance between the starting point of the chain can be calculated, so that:

$$<r_c^2> = Nl^2 \tag{2.4}$$

where l is the segment length. This is for a *freely jointed chain* and no account has been taken of finite bond angles, or the excluded volume interactions of both neighbouring segments and distant segments along the chain that interact as the 'walk' takes them back to cross the chain. For any real polymer chain there are fixed bond angles, and rotation around the bonds is markedly reduced if bulky side groups are present, and so the 'walk' is much more spatially extended. In other words, a real chain is much stiffer than a freely jointed one and the conformation is expanded, with the mean-square end-to-end distance being expressed as follows:

$$<r_c^2> = c_\infty Nl^2 \tag{2.5}$$

Here, c_∞ is the 'characteristic ratio', with some typical values being given in Table 2.1.

Table 2.1. Values of the characteristic ratio for various polymers

Polymer	c_∞
Freely jointed chain	1
Tetrahedral bond angle	2
Poly(ethylene oxide)	4.1
Polydimethylsiloxane	5.2
Poly(12-hydroxystearic acid)	6.1
Polystyrene	9.5

The solvent chosen to dissolve the polymer is also important. In a *good* solvent, a chain segment is surrounded by the maximum number of solvent molecules that can be packed around it. In a *poor* solvent, there is an increased probability of there being other chain segments around any particular segment. (Thus, as the quality of the solvent decreases, the polymer chains become insoluble.) The conformation of a polymer chain in solution is a spheroidal coil which can be characterized by the root-mean-square end-to-end chain dimension, as follows:

$$<r_c^2>^{0.5} = (c_\infty N)^{0.5} l \quad \text{Ideal solution}$$
$$<r_c^2>^{0.5} = (c_\infty N)^{0.6} l \quad \text{Good solvent} \quad (2.6)$$

We may also characterize the coil dimension in terms of its radius of gyration, R_g. This is the average distance of the polymer segments from the centre of mass of the coil:

$$R_g^2 = \frac{\sum_i m_i r_i^2}{\sum_i m_i} = \frac{<r_c^2>}{6} \quad (2.7)$$

The radius of gyration can be experimentally obtained from light scattering measurements of dilute polymer solutions. It is interesting to compare the dimensions of a polymer molecule dissolved in a good solvent with what we would expect from the bulk density of the same polymer (Figure 2.1). We can see from this figure that the dimensions of the molecule in solution are very much greater than they would be in the dry state and so the concentration of polymer within the coil in solution is very low indeed. It is important to keep in mind that the 'connectivity' along the chain demands this very open structure.

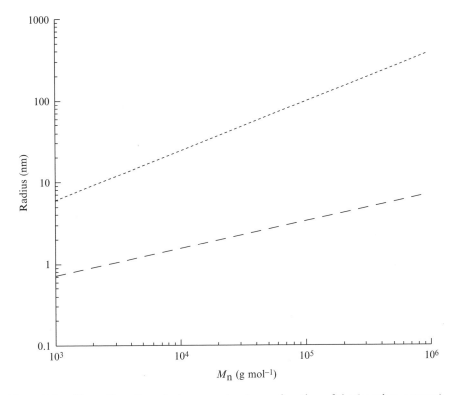

Figure 2.1. The radius of a polystyrene molecule as a function of the (number-average) molecular weight: (- - -) radius of gyration in a good solvent; (– – –) radius of an amorphous sphere.

3.1 The Gaussian Chain

At first sight, it may appear that the model of a freely jointed chain would not be a good picture of a real polymer with rigid bond angles, even though an expansion factor has been included. However, if groups of several bonds are considered, the co-operative effect is to add flexibility. Thus, the artifice is to consider the chain segments as a larger unit which could contain, for example, five bonds. This allows the flexibility to be reintroduced. The complication is that the bond length is now a variable and so our random walk no longer has the constraint of equal step lengths. This is now closer to the diffusion problem and the result is that there is a Gaussian distribution of step lengths. Figure 2.2 presents a schematic illustration of part of a chain. The number of segments in this example is $N/5$, with the mean step length from the Gaussian distribution as l', while $l'/(l \times 5^{0.6})$ is taken care of in the value of c_∞ that we use.

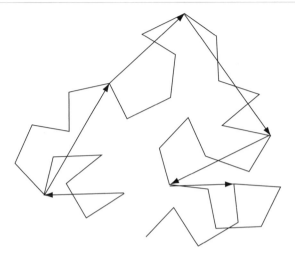

Figure 2.2. Schematic of a three-dimensional random walk with fixed bond angles and fixed step lengths. The arrows indicate the increased flexibility introduced by designating sections containing five bonds as one chain segment at the cost of variable step lengths.

4   THE FLORY–HUGGINS THEORY OF POLYMER SOLUTIONS

This theory [2, 3, 7] describes the thermodynamics of polymer solutions in sufficient detail for the purposes of our present text. The starting point is a simple lattice model. A cubic lattice is normally used, as this is easy to visualize and there is no real gain in understanding from using other lattice structures such as a tetrahedral one. The key feature of this lattice is that the solvent molecules are assumed to be the same size as the segments of the polymer chain. The entropy of mixing is estimated from the number of possible configurations on the lattice and the enthalpy from the interactions between the various components.

4.1   The Entropy of Mixing

This is calculated from the number of ways that a polymer molecule can occupy the sites on a filled lattice. Sites not occupied by chain segments must be occupied by solvent molecules. Thus, there are n_s lattice sites occupied by solvent molecules with n_p sites occupied by polymer chains (illustrated in Figure 2.3). The result is as follows:

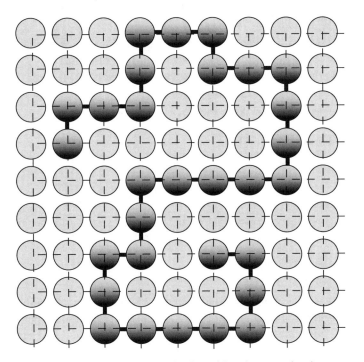

Figure 2.3. Illustration of a filled cubic lattice with solvent molecules surrounding a polymer chain. Note that the lattice must be fully occupied and the solvent molecular size is equal to the chain segment size.

$$\Delta S_{\text{mix}} = -k_{\text{B}}(n_{\text{s}}\ln \varphi_{\text{s}} + n_{\text{p}}\ln \varphi_{\text{p}}) \tag{2.8}$$

The volume fractions, φ, are given by the following:

$$\varphi_{\text{s}} = \frac{n_{\text{s}}}{n_{\text{s}} + Nn_{\text{p}}}; \quad \varphi_{\text{p}} = \frac{Nn_{\text{p}}}{n_{\text{s}} + Nn_{\text{p}}}$$

$$\frac{\varphi_{\text{s}}}{\varphi_{\text{p}}} = \frac{n_{\text{s}}}{Nn_{\text{p}}}, \text{ and so } n_{\text{s}}\varphi_{\text{p}} = Nn_{\text{p}}\varphi_{\text{s}} \tag{2.9}$$

This is analogous to the problem of mixing two liquids except that volume fractions have been used instead of mole fractions. The mole fraction of a high-molecular-weight polymer in solution would always be very small, even when the volume fraction is large, due to that high molecular weight. (If the segments of the chain were not connected, we would just be mixing two miscible liquids and $\Delta S_{\text{mix}} = -k_{\text{B}}(n_1\ln x_2 + n_2\ln x_1)$ where x is the mole fraction of either the solvent, 1, or the solute, 2.)

4.2 The Enthalpy of Mixing

Flory [2, 3, 7] calculated this by considering the local energy changes as we mix polymer segments with solvent, as shown in Figure 2.4.

Therefore, the change in the internal energy is given by:

$$\Delta U_{sp} = U_{sp} - \frac{U_{ss} - U_{pp}}{2} \qquad (2.10)$$

and the enthalpy of mixing by:

$$\Delta H_{mix} = \Delta U_{mix} + P\Delta V$$
$$\Delta H_{mix} = N_c \Delta U_{sp} + \sim 0 \qquad (2.11)$$

The 'PV' term can be approximated to zero, and so the number of contacts per unit volume, N_c, is as follows:

$$N_c = \varphi_s n_p N z \qquad (2.12)$$

i.e. the number of contacts is equal to the probability of a site being occupied by a solvent molecule, φ_s, multiplied by the number of polymer segments, $n_p N$, and a coordination number for the lattice contacts, z. Therefore, the enthalpy of mixing becomes (with Equation 2.9):

$$\Delta H_{mix} = \varphi_p n_s \chi k_B T \qquad (2.13)$$

where we have defined χ, which is known as the *Flory–Huggins interaction parameter*, as the internal energy change per segment on mixing relative to thermal energy as:

$$\chi = \frac{\Delta U_{sp} z}{k_B T} \qquad (2.14)$$

We may now write the free energy of mixing as follows:

$$\Delta G_{mix} = \Delta H_{mix} - T\Delta S_{mix}$$
$$\Delta G_{mix} = k_B T(n_s \ln \varphi_s + n_p \ln \varphi_p + n_s \varphi_p \chi)$$
$$\Delta G_{mix} = RT(N_s \ln \varphi_s + N_p \ln \varphi_p + N_s \varphi_p \chi) \qquad (2.15)$$

solvent solvent   segment segment   2(solvent segment)

Figure 2.4. Interactions occurring on mixing polymer segments with solvent.

where N_s, etc. are in molar quantities. Formally, Equation (2.10) should have used a free energy component, so χ really contains an entropy term; however, as this is determined experimentally we do not generate practical problems by this approximation. For a polymer in a good solvent, the value of χ is found to be between 0.5 and 0.1. Now that we have the free energy of mixing as a function of the solution concentration, we may calculate the osmotic pressure of the dilute polymer solution as follows [2]:

$$\Pi = \frac{RT}{\bar{v}_s} \left[\frac{\varphi_p}{N} + \left(\frac{1}{2} - \chi \right) \varphi_p^2 + \ldots \right] \tag{2.16}$$

with $\bar{v}_s$ as the molar volume of a solvent molecule. The significance of the χ-parameter can be immediately appreciated from Equation (2.16). When we have the condition that $\chi = 0.5$, the polymer/polymer interaction term vanishes and the osmotic pressure for the dilute solution and the osmotic pressure relationship is similar to the van't Hoff expression. Using the relationships from the lattice model, Equation (2.16) can be recast in the more familiar form:

$$\frac{\Pi}{c_p} = RT \left[\frac{1}{M_n} + \left(\frac{1}{2} - \chi \right) \left(\frac{\bar{v}_p N}{M_n^2} \right) c_p + \ldots \right] \tag{2.17}$$

with c_p as the polymer concentration in mass per unit volume and $\bar{v}_p$ as the molar volume of the polymer. The coefficient of the polymer concentration on the right-hand side of Equation (2.17) is the osmotic second virial coefficient, B_2, and is the slope of the curve of the reduced osmotic pressure as a function of concentration, as illustrated in Figure 2.5. As the solvent properties are changed, by changing the temperature, pressure or composition, for example, the value of χ changes and the 'quality' of the solvent can be defined as follows:

(1) $\chi < 0.5$ – we have a 'good' solvent for the polymer;
(2) $\chi \sim 0.5$ – the solvent is termed a θ-solvent;
(3) $\chi > 0.5$ – the solvent is a 'poor' solvent, and as the value increases much above 0.5, the polymer solubility reduces, even though it may be swollen by the solvent.

For example, polystyrene is soluble in cyclohexane. The θ-temperature is 38.5 °C, and so at 45 °C cyclohexane is a good solvent for polystyrene. At the θ-temperature, the conformation of the polymer molecule is minimally disturbed by solvent–chain segment interactions and is as close to a random coil as obtainable by that molecule.

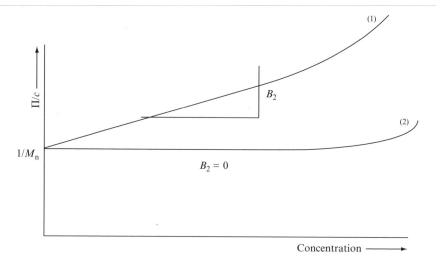

Figure 2.5. Schematic curves for the reduced osmotic pressure as a function of concentration for the conditions (1) $\chi < 0.5$, and (2) $\chi = 0.5$.

5 POLYMER SOLUTION PHASE BEHAVIOUR

When we consider the solubility of a solute in a solvent, our normal experience is for the solubility to increase as the temperature is increased and conversely, if we cool a solution, at some temperature we will observe the solute phase coming out of solution. This is the usual pattern with polymers in a good solvent. When the system is cooled below the θ-temperature, the solvent becomes progressiveley poorer and two phases will be observed with the polymer-rich phase being polymer swollen with solvent. The phase boundary is known as the *upper consolute solution temperature* (UCST). Above this temperature, a single phase is formed. In many aqueous systems, and occasionally in some polar organic solutions of polymer, another phase boundary – at the *lower consolute solution temperature* (LCST) – can be found where phase separation can occurs as the solution is heated. Water-soluble polymers contain polar groups such as hydroxyl, carboxylic acid or ether groups which can take part in the hydrogen (H)-bonding structure of water. As the temperature increases, the H-bonding is reduced, and the polymer ceases to be in a good solvent and phase separation can occur. A general solubility diagram is presented in Figure 2.6. This type of behaviour is also observed with non-ionic surfactants in aqueous solution, with the LCST being termed the *cloud point*.

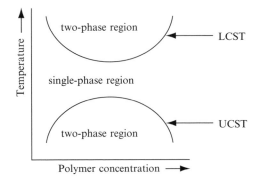

Figure 2.6. Illustration of a general solubility diagram: LCST, lower consolute solution temperature; UCST, upper consolute solution temperature.

However, because of the connectivity of the groups making up a high polymer, there are further subtleties in the phase behaviour of polymer solutions. Consider a polymer in a θ-solvent. In dilute solution, the polymer is as close to a random coil configuration as it can obtain. As the concentration is increased, the polymer molecules interpenetrate extensively as the interactions between polymer segments, polymer/solvent and solvent molecules have similar energies in a 'θ-condition'. When the polymer is in a 'better than θ-solvent', the situation changes somewhat. At low concentrations (the dilute regime illustrated in Figure 2.7(a)), the polymer coils are in an expanded configuration and, on average, are separated from each other. Hence, if we were to measure the concentration profile across a section of solution there would be clear variations, as illustrated schematically in Figure 2.7(b). As the concentration increases to a value denoted by c^*, the polymer coils become 'space-filling'. The global polymer concentration is just equal to that which would be calculated for a single coil, and so by using Equation (2.6) we have the following:

$$c^* \sim \frac{M_n}{N_A <r_c>^{3/2}} = \frac{M_n}{N_A (c_\infty N)^{9/5} l^3} \qquad (2.18)$$

As the solution concentration increases, the variation in local concentration becomes small as interpenetration increases, and the polymer solution is said to be 'concentrated'. This is illustrated in the lowest part of Figure 2.7(b). (c^{**} is the concentration where the individual coils are no longer discernible and the chains are in their 'ideal' state). This boundary moves to higher concentrations as the temperature is increased to above the θ-temperature. The excluded volume interactions in the concentrated state result in the osmotic

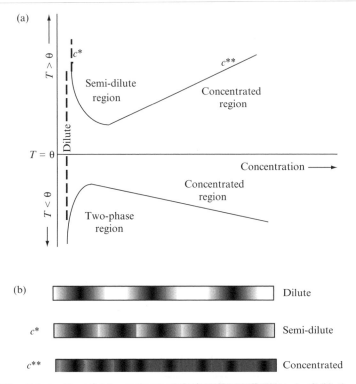

Figure 2.7. Schematics of (a) a polymer solution phase diagram, and (b) the concentration variation across a polymer solution, showing the boundary from dilute to semi-dilute as the coils 'space-fill' at c^* and the approach to a uniform concentration at c^{**}.

pressure continuing to increase with increasing polymer concentration, as well as decreasing the diffusivity of the molecules.

In both the semi-dilute and concentrated regimes, each polymer molecule is a component of a mesh due to the interpenetration of each molecule by its neighbours. The mesh size is referred to as the *correlation length* and decreases with increasing concentration until the dimension is of the order of the segment size in the melt state. Structural relaxation of the bulk system is effected by the diffusion of the molecules. When they are part of an entangled mesh, the net motion is by the wriggling or *reptation* [5] of each chain through the mesh. The model for this motion [5, 6] is of a chain moving through a tube. The dimensions of the tube cross-section are governed by the mesh size as the walls are formed by the surrounding molecules. Of course, as the concentration increases and the mesh size is reduced, the dynamics are slowed.

6 POLYMERS AT SURFACES

The starting point is to consider the interaction energy between the atoms or molecules making up this third component with those between the solvent and macromolecular species. If we use the concept of the Flory χ-parameter, then we may assign a value for the interaction with the surface by considering the interaction energies between the polymer/solvent, polymer/surface and solvent/surface. So, if the value of $\chi < \chi_{surf}$, the polymer will not absorb (where χ_{surf} is the polymer/surface value). Conversely, if the value of $\chi > \chi_{surf}$, the polymer will adsorb. Detailed modelling has been carried out by Scheutjens and Fleer [8], who used the lattice model at a surface and varied the χ-parameter over the first few layers. This enabled predictions of concentration profiles to be made for both adsorbed homopolymers and adsorbed copolymers. The profile has also been modelled as a 'self-similar mesh' by deGennes [9, 10]. The details of the outer part of the concentration profile become of interest in the discussion of particles stabilized by adsorbed macromolecules. Figure 2.8 illustrates the concentration profile for a non-adsorbing polymer. To obtain a uniform polymer concentration right up to the interface, the conformations in the different parts of the polymer would have to be reduced as that part of the coil close to the surface becomes more concentrated. This is energetically unfavourable without a competing attraction from the interface and the result is a *depletion layer* where the local concentration is lower than the global average within a distance of $\sim R_g$ away from the surface. When the enthalpy change for the adsorption, coupled with the

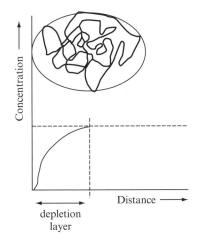

Figure 2.8. Illustration of the closest approach of a non-adsorbing polymer coil to a surface, showing the reduction in the local polymer concentration close to the surface from the average value in the solution – the latter is termed the *depletion layer*.

increase in entropy of solvent molecules displaced from the surface, is greater than the decrease in entropy due to the restriction on polymer conformation, the free energy is favourable for adsorption and the polymer will stick to the surface. Figure 2.9 illustrates the type of conformation that occurs for a polymer adsorbed from a θ-, or better, solvent. In a poor solvent, of course, the polymer will be adsorbed in a dense layer on the surface. Figure 2.10 shows the concentration profiles in the surface layer. Note that the tails project further into the solution phase than the loops and so the total concentration profile falls to that of the tails at the outer periphery.

Homopolymers are not usually added to colloidal systems to enhance the colloidal stability by adsorption. They are, however, frequently added as *rheology*

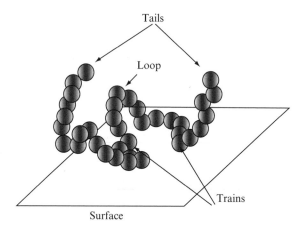

Figure 2.9.   Representation of the conformation of a polymer adsorbed at an interface, showing the features of 'tails', 'loops' and 'trains'.

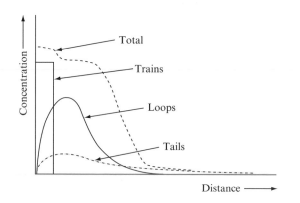

Figure 2.10.   Illustration of the concentration profile of an adsorbed polymer.

modifiers (thickeners). Copolymers are much better candidates to act as stabilizers. The molecular design is chosen to be amphipathic so that part of the molecule is in a 'worse than θ-solvent', while the rest is in a 'better than θ-solvent'. This provides strong attachment to the surface while maximizing the extension of the soluble moieties. A variety of copolymer configurations are available and are shown in Figure 2.11. Poly(vinyl alcohol) is prepared by the partial hydrolysis (e.g. ∼ 80 %) of poly(vinyl acetate) and is a commonly used random block copolymer stabilizer. The surface configuration adopted by such random block copolymers will be similar to that illustrated in Figure 2.9.

The adsorption isotherms measured for adsorbing polymers are usually of the high-affinity type so that when low amounts of polymer are added this is all adsorbed. Figure 2.12 illustrates the type of curve frequently obtained in

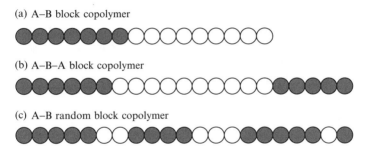

(a) A–B block copolymer

(b) A–B–A block copolymer

(c) A–B random block copolymer

Figure 2.11. Some examples of block copolymers: (a) A–B, e.g. poly(propylene oxide)-*co*-poly(ethylene oxide); (b) A–B–A, e.g. poly(12-hydroxystearic acid)-*co*-poly(ethylene oxide); (c) A–B random, e.g. poly(vinyl alcohol).

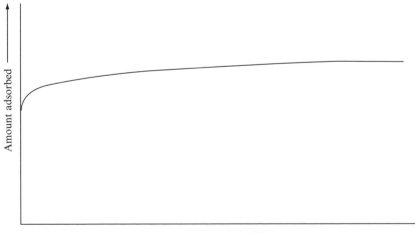

Figure 2.12. Illustration of the type of high-affinity isotherm found for most adsorbing polymers used for stabilizing colloidal particles.

this case. The determination of an isotherm is usually carried out by equilibrating aliquots of a dispersion with varying amounts of polymer, separating the particles and carrying out an assay for the polymer in solution. This has some practical difficulties, however. For example, the equilibration time can be quite long – in some cases, it can be days. The lower-molecular-weight species diffuse more rapidly and can occupy the surface first. Subsequent displacement by larger molecules may then occur but this is slow due to the 'multi-point' attachment of the chains with the added complication of slow surface motion to optimize the conformation. This behaviour makes the mixing process important as the desorption and reorganization of polymer is very slow. An added problem is that the equilibrium concentration of polymer in the solution phase, close to the onset of full coverage, is low, thus making the analysis difficult. When we use the dimensions of a polymer coil given in Figure 2.1 above, and pack these together on a surface, monolayer coverage corresponds to a value of the order of $1\,\mathrm{mg\,m^{-2}}$. With the specific surface area for many colloidal systems lying in the range 1 to $100\,\mathrm{m^2\,g^{-1}}$, it is clear from this that quite small amounts of polymer are sufficient to saturate the surface of a dispersion.

7   POLYMER CHARACTERIZATION

The molecular weight is a key piece of information that we require for any polymer system. There are several experimental options available to give us such information. It is always useful to employ two different methods as this will provide an insight into the polydispersity of the polymers that we are using. At this point, it is worthwhile to briefly review the methods most frequently used.

7.1   Intrinsic Viscosity

It is straightforward to obtain the viscosity of a polymer solution in a simple U-tube viscometer where the time is measured for a known volume of solution to flow through a capillary tube [11]. The viscosity of a polymer solution is described by the Huggins equation, as follows:

$$\eta = \eta_0[1 + [\eta]c + k_H[\eta]^2c^2 + \cdots]\qquad(2.19)$$

where c is the concentration of the polymer solution with a viscosity of η, η_0 is the solvent viscosity, $[\eta]$ is the *intrinsic viscosity*, and k_H is the Huggin's coefficient. The intrinsic viscosity is directly related to the size and shape of the molecule in solution and therefore to the molecular weight, while the

Huggin's coefficient is a function of the pair-wise interaction of the molecules. This equation could be extended to include higher-order interactions, but as written is the low-concentration result. Equation (2.19) can be rearranged to give the following linear expression:

$$\frac{\left(\frac{\eta}{\eta_0} - 1\right)}{c} = [\eta] + k_H[\eta]^2 c \qquad (2.20)$$

The term on the left-hand side of Equation (2.20) is known as the *reduced viscosity*. A plot of the reduced viscosity versus the solution concentration gives a straight line with the intrinsic viscosity as the intercept. Experimentally, the problem is that at low concentrations the relative viscosity, (η/η_0), is very close to unity so when we calculate the reduced viscosity we need data of high precision. For example, if our viscometer had a flow time of 120 s, we would like data to be reproducible to within 0.2 s. This requires a great deal of care, cleanliness and excellent temperature control as the viscosity changes exponentially with temperature.

The molecular weight can be obtained from the intrinsic viscosity by using the Mark–Howink Equation:

$$[\eta] = K M_v^a \qquad (2.21)$$

where K and a are constants for a given polymer and solvent pair. The value of a is dependent on the quality of the solvent. In a θ-solvent, $a = 0.5$ and in 'better than θ conditions' it can rise to 0.8. There is a comprehensive set of data available in the literature [12]. Equation (2.21) was originally derived by using polymers with narrow molecular weight fractions but, of course, with the broad distribution that we usually work with, the value is an average. The viscosity is a function of both the number density and the size in solution and the average is in between the value we would calculate on the basis of number and the average on the basis of weight.

7.2 Limiting Osmotic Pressure

The osmotic pressure–concentration relationship was given in Equation (2.17) while the measurement of osmotic pressure provides a useful method of determining the molecular weight. The osmotic pressures of a series of solutions of different polymer content are measured and a linear plot of Π/c as a function of c should be obtained at low concentrations, as illustrated in Figure 2.13. The number-average molecular weight is obtained from the intercept and the value of the χ-parameter from the slope.

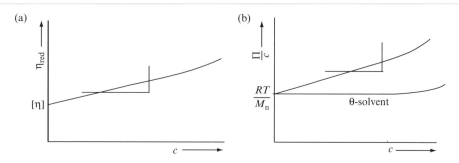

Figure 2.13. (a) The reduced viscosity plotted as a function of concentration for a polymer solution. The intercept yields the intrinsic viscosity and the slope is $k_H[\eta]^2$. (b) The reduced osmotic pressure plotted as a function of concentration for a polymer solution. The number-average molecular weight is obtained from the intercept and the 2nd virial coefficient from the slope, with the latter reducing to zero as χ approaches 0.5.

The type of osmometer used is dependent on the molecular weight of the polymer. A membrane osmometer is used for $M_n > 10^4$ Da, while lower-molecular-weight polymers are characterized by using a vapour pressure osmometer. With a membrane osmometer, pressure is applied to the solution to prevent the transfer of solvent across a microporous membrane. The vapour pressure instrument measures the change in temperature of an evaporating drop on the tip of a syringe. The higher the vapour pressure, then the faster is the evaporative cooling rate. The vapour pressure of a solution is a function of the concentration and the molecular weight of the solute. The equipment is calbrated with standard polymers of narrow molecular weight distributions.

7.3 Angular Light Scattering

When light passes through a medium and penetrates a region where the refractive index changes, light will be scattered radially from that region. A dissolved polymer molecule is one such region and each molecule will scatter light. The scattering of electromagnetic radiation is discussed in some detail later in this volume. However, at this point we should recognize that the amount of light scattered is a function of the number of scattering units per unit volume, their size, and the difference in refractive index of the scattering unit and the surrounding medium.

The simplest approach is to treat the scattering molecules as 'point scatterers'. As the dimension in solution has R_g as less than 5 % of the wavelength, this is a good approximation. We measure the relative intensity of light scattered at a given angle to give the *Rayleigh ratio*, $R(\Theta)$, as follows:

$$R(\Theta) = \frac{Ir^2}{I_0 v_s} \tag{2.22}$$

where I is the transmitted intensity, I_0 is the incident intensity, r is the path length from the cell to the dectector and v_s is the scattering volume. It is convenient to measure the scattering at $90°$ and we can express the dependence of the Rayleigh ratio on the concentration of a dilute polymer solution as follows [13, 14]:

$$\frac{K_1 c}{R(90)} = \frac{1}{M_w} + 2B_1 c \tag{2.23}$$

with the optical constant:

$$K_1 = \frac{2\pi^2 n_0^2}{\lambda^4 N_A} \left(\frac{\partial n}{\partial c}\right)^2 \tag{2.24}$$

being dependent on the incident wavelength, λ, the refractive index of the solvent, n_0, and the rate of change of refractive index with polymer concentration; B_1 is the second virial coefficient and is dependent on the quality of the solvent. In this case, the weight-average molecular weight is obtained from the intercept of a plot of Equation (2.23) at zero concentration.

7.4 Gel Permeation Chromatography

When a polymer solution is passed slowly through a porous bed, or swollen cross-linked gel, diffusion of the macromolecules means that they explore the pores as they pass through. This slows the passage of the macromolecules as when inside a pore, they are out of the flow and only pick up forward motion when they emerge. Smaller molecules can explore small pores as well as large ones with the result that they spend more time inside the pores. The net result is that a separation occurs if the column is long enough. The larger macromolecules emerge first, with the smaller fraction emerging last as the column is eluted. The concentration of the polymer in the eluent is measured. A common and convenient method for this is to measure the refractive index of the solution as it passes through. This provides a continuous electronic read-out. The concentration can be calculated from a calibration curve of the refractive-index dependence of solutions of the polymer being studied. The residence time on the column is calibrated by using narrow-molecular-weight-distribution standards. This technique provides a molecular weight distribution and not just an average value. In addition, the method may be scaled-up to provide a preparative route to small amounts of narrow-distribution material.

Columns are available for both aqueous or non-aqueous systems. Adsorption of the macromolecules onto a chosen column material can prevent the technique from being used successfully, and so the choice of an appropriate packing material and solvent system is important.

The characterization of copolymers such as A–B–A block or hydrophobically modified polymers can present special difficulties. Such polymers are synthesized so that part of the macromolecule is in a poor solvent. The result is that some self-assembly may occur and lead to the measurement of aggregates of macromolecules. Characterization of molecular weight needs to be carried out in a solvent system which will suppress such aggregation. It must then be recognized that the conformation in the solvent system used for the final application may require some additional experimental work.

8   SURFACTANTS IN SOLUTION

A *surfactant*, or *surface active agent*, is a general term used to describe molecules that interact with an interface. These consist of two parts, one of which is highly soluble in one of the phases while the other is not. They are small mobile molecules which are widely used in colloidal systems. For example, they are used as soaps, detergents, dispersants, wetting agents and germicides. Their structures consist of a hydrophobic tail which is usually a hydrocarbon, although fluorocarbon and dimethylsiloxane chains can be used, with a polar hydrophilic head group which may be ionic or non-ionic. This type of molecular structure gives rise to the non-ideality of solutions of surfactants and their phase behaviour. Figure 2.14 shows some examples of different types of surfactant molecules. Surfactants are used in both aqueous and non-aqueous systems and, although we usually think of the synthetic materials that are manufactured in large quantities, there are some very important naturally occurring ones. For example, the surfactants present in our lungs are vital for their operation, as are the bile salts produced by the pancreas, and act to disperse dietary fat into colloidal size droplets (or *chylomicra*) which pass into the blood stream where they are utilized by the body. The phospholipid lecithin is a constituent of cellular membranes. The fatty acids are also surfactants and form the source of soaps, the manufacture of which consists of basically producing the sodium salt in a high-concentration phase which can be conveniently handled.

8.1   Dilute Solutions

We will mainly concern ourselves with aqueous solutions but it should be kept in mind that analogous behaviours may be found in other solvents. As surfactant is added to water, the molecules dissolve. In most cases, an

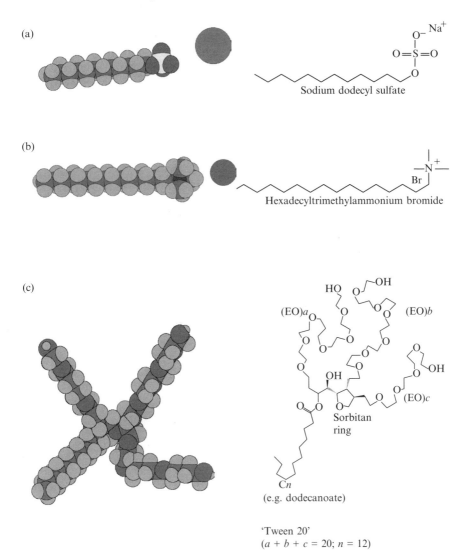

(a)

Sodium dodecyl sulfate

(b)

Hexadecyltrimethylammonium bromide

(c)

(EO)a

(EO)b

(EO)c

Sorbitan
ring

Cn
(e.g. dodecanoate)

'Tween 20'
($a + b + c = 20$; $n = 12$)

Figure 2.14. (Plate 1). Some examples of different types of surfactants: (a) anionic; (b) cationic; (c) non-ionic (EO, ethylene oxide).

increase in temperature aids this process but this is not universally true. In solution, the polar head groups are hydrated – they can take part in the H-bonding structure of the water. The hydrocarbon tails, on the other hand, cannot do this and we can visualize a discontinuity in the structure of the water around these tails. This has been referred to as a 'cage' and is cited as the origin of the 'hydrophobic effect'. What this means is that for the tail to be in solution there is a increase in free energy relative to the reduction in the number of H-bonds, which we may think of as proportional to the area of the 'cage'. If several of these tails are brought together into an aggregate, the surface area that is required to surround them would be less than the sum of the areas of the individual 'cages'. Furthermore, there will be an additional enthalpic term from the van der Waals attraction between the tails. Opposing this, of course, is the decrease in entropy associated with the clustering of the chains, although the entropic term for the water increases. With ionic head groups, the electrostatics oppose their close approach, but counter-ion binding reduces this. In addition, water is also partially freed from the hydration sheath of the head groups. Any interface, such as the air–water or oil–water interface, also provides an opportunity for the tails to move out of the water structure.

With this picture in mind, let us consider the process of adding surfactant to water in increasing amounts and changing the temperature in each case. What we are doing, of course, is investigating the phase behaviour from a series of *isopleths* (varying temperature at fixed concentrations). Figure 2.15 presents the type of phase diagram which we could expect. At low temperatures, the solubility is low with surfactant molecules in equilibrium with the surfactant solution. There is a critical point known as the *Krafft point* [15, 16], and at temperatures higher than the value at this point (the *Krafft temperature*) the solubility appears to increase rapidly and the solution phase consists of surfactant aggregates or *micelles*, as well as single molecules. Below the Krafft temperature, micelles are not formed. The concentration at which micelles are produced is known as the critical micelle concentration (cmc). The latter varies with temperature above the Krafft temperature.

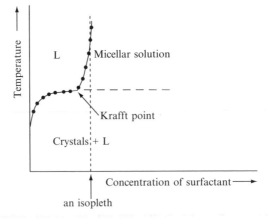

Figure 2.15. Schematic phase diagram for a dilute surfactant–water system: L represents a 'simple' solution. The Krafft point is the critical point, and above the Krafft temperature the solution consists of surfactant micelles as well as surfactant molecules.

8.2 Micellization

Provided that we keep our solution above the Krafft temperature for our surfactant, there is clear evidence for a sharp self-assembly process occurring at a particular concentration from a distinct change in the physical properties of the surfactant solution. If the surface tension is measured for increasing concentrations of surfactant in solution, the former decreases steadily as increasing adsorption of surfactant molecules at the air/water interface disrupts the local H-bonding and makes the surface more 'oil-like'. The extent of the adsorption can be calculated from the Gibbs adsorption equation, as discussed in Chapter 1. However, quite suddenly, the slope of the surface tension versus concentration curve decreases to almost zero. This indicates that the adsorption at the interface is now constant, although our solution concentration is increasing. One explanation that might occur to us is that the surface is now saturated by a monolayer and so no more sites are available. However, then we find that other properties also change at the same concentration. For example, the rate of increase in osmotic pressure fall dramatically to a plateau. Light scattering experiments show a sharp increase in turbidity. These latter two observations suggest that the increased amount of surfactant that we have added is not in the solution phase. Measurements of the equivalent conductance with increasing concentration show a marked decrease in slope after the critical micelle concentration, so indicating that we now have much less mobile charged units than we would expect from the individual molecules of surfactant. Figure 2.16 illustrates the type of change that we observe. Other measurements, e.g. by using NMR spectroscopy, also indicate marked changes at the same concentration. All of these experiments provide clear evidence of a phase change but with the formation of a sub-microscopic phase. When we measure the size of the units of the new phase by light scattering or neutron scattering, for example, this size is twice the length of the linear surfactant molecule.

As the surfactant concentration is increased to a level a little above the cmc, we have spherical units – micelles – with a diameter twice that of the individual molecules, in addition to a constant concentration of single molecules. Attention was drawn above to the concept of the 'hydrophobic effect' and we may think about the micelle in these terms. Consider some typical anionic surfactants such as sodium dodecanoate or sodium dodecyl sulfate. In both cases, there is a 12-carbon paraffin chain with a polar head group. The head groups will remain in the water phase, with the tails in the spherical 'oily' phase. If the radius is larger than a stretched surfactant molecule, some of the heads would automatically have to be buried in the oil phase, which would be energetically expensive. At any given moment, some of the tails will be linear, while others will be bent to fill the volume to give a density similar to that of the bulk paraffin. With a CH_2-CH_2 distance of 0.127 nm, we can easily

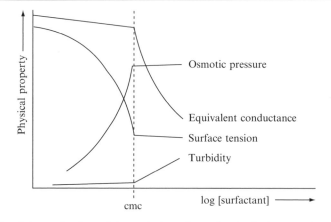

Figure 2.16. Schematic plots of the variation of certain physical properties of a surfactant solution at the critical micelle concentration.

calculate that there would be 60 or 80 molecules per micelle for the dodecanoate or sulfate molecules, respectively (the latter having a longer head group). If we take the diameter of a paraffin chain as 0.4 nm, we can compare the 'water contact area' for the appropriate number of chains with that of the outer surface of the micelle. Of course, we need to allow for the area occupied by the polar head groups on the surface (0.2 nm² per COOH group, for example). When this is done, we find that about two thirds of the micellar surface consists of carbon chains in contact with water, although this is only one sixth of the area that would be in contact if the molecules were separated. In addition to increasing the entropy of the water by freeing up the local structure, there will also be a contribution from the hydrocarbon because its motion is less constrained than when restricted by the water 'cage'. A detailed thermodynamic treatment would include contributions from the mutual repulsion of the head groups mitigated by the ion atmosphere around the micelle.

Of course, the question is 'whether there is a variation in the size of micelles around a mean value?'. To answer this, we need to consider the aggregation process in a little more detail. Aggregation is a progressive process which can be represented by the following multiple equilibria:

$$S_1 + S_1 \rightleftharpoons S_2$$
$$S_1 + S_2 \rightleftharpoons S_3$$
$$\downarrow$$
$$S_1 + S_{i-1} \rightleftharpoons S_i, \text{ etc.} \tag{2.25}$$

which can be represented as follows:

$$iS_1 = S_i \tag{2.26}$$

where we have a single molecule, S_1, forming a micellar aggregate, S_i. The equilibrium constant for this 'reaction' is given by the following:

$$K_i = \frac{x_i}{(x_1)^i} \tag{2.27}$$

where x_1 and x_i are the mole fractions of the monomers and micelles, respectively. The equilibrium constant gives the standard free energy of formation of the micelle containing i monomers as follows:

$$\Delta G^0(i) = -RT\ln K_i = -RT\ln x_i + iRT\ln x_i \tag{2.28a}$$

that is:

$$\frac{\Delta G^0(i)}{RT} = i\ln x_1 - \ln x_i \tag{2.28b}$$

The standard free energy of formation of a micelle has contributions from the following:

(a) a decrease as the hydrophobe–hydrophobe interactions replace the hydrophobe–water interactions;
(b) an increase as we form an interface between the micro-phase of the micelle and the surrounding water;
(c) an increase as we bring the hydrophilic head groups closer together by concentrating the hydrophobes in one place.

Everett [17] has suggested the following equation for the standard free energy of formation based on these contributions:

$$\frac{\Delta G^0(i)}{RT} = -a(i-1) + b(i-1)^{2/3} + c(i-1)^{4/3} \tag{2.29a}$$

where the coefficients, a, b and c, have their origins in the contributions listed above, with values that will vary with the chemical architecture of the system For example, a will become increasingly negative as the chain length of the hydrophobe increases and c would increase if the molecule had a charged hydrophilic group To illustrate the implications of Equation (2.29a), we can rewrite it with suitable numerical values for these three parameters, as follows:

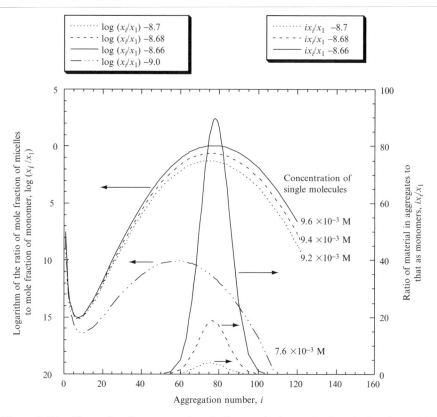

Figure 2.17. The ratio of monomer in micelles to single species in solution for a C12 surfactant.

$$\frac{\Delta G^0(i)}{RT} = -18.8(i-1) + 19.55(i-1)^{2/3} + 1.25(i-1)^{4/3} \qquad (2.29b)$$

By combining Equation (2.29b) with Equation (2.27), we can compute the ratio of the mole fractions of the micelles to that of the monomers as a function of i. By using single molecule concentrations at $\sim 10^{-2}$ M (which is a typical value for the critical micelle concentration of a surfactant), we see from Figure 2.17 that there is a maximum in the curve at a value of $i \sim 80$. However, the concentration has to get very close to the cmc before there is a significant proportion of material in the aggregates (the numerical value of $\log(x_i/x_1)$ must approach 0). If the concentration falls to around two thirds of the cmc, the maximum disappears, thus indicating that aggregation would be absent.

As the concentration is increased to the cmc and a little above, we find that the majority of the surfactant is present in aggregates and that these are of a preferred size. When we express the data as the ratio of the material in a micelle with an aggregation number i to the monomer level as a function of that aggregation number, it becomes quite clear that a preferred micellar size is obtained with a sharp distribution around the modal value. We can conclude that at concentrations a little above the critical micelle concentration we have a phase-separation process which gives a monodisperse or uniform-size micro-phase. It is the balance between the opposing molecular interactions that define the preferred size and prevents the separating phase growing indefinitely. With ionic surfactants, the counter-ion type and charge play a part and the micelle has to be modelled by taking into account the potential distribution around the sphere [18]. The value of the critical micelle concentration will be a function of the electrolyte concentration as well as the temperature [19].

The micelles are not static, of course – the chains and head groups are mobile. Bound counter-ions of ionic head groups, which neutralize 30–50 % of the charge, are not associated with a particular group but are mobile across the surface. Surfactant molecules can leave and rejoin micelles in a dynamic equilibrium, with the residence time being of the order of 1 µs. In addition, we should not think of a micelle as being a permanent entity but a 'unit' with a lifetime of the order of 1 ms. Even when there are oil-soluble molecules dissolved in the interior of the micelle (this is known as *solubilization*), the micelle is still not a permanent entity. The phenomenon of solubilization will only extend the lifetime. Quite large molecules such as naphthalenic dyes can be solubilized by micelles and this process causes the micelle to increase a little in size. A typical micelle would only solubilize about two naphthalene molecules and so the size increase is not great. Aliphatic alcohols with chain lengths of six to twelve carbon atoms also stabilize the micelle by reducing the interactions between head groups. Figure 2.18 shows the variation of the cmc with chain length for a number of different surfactants.

The above picture illustrates the phase behaviour at concentrations close to the cmc. At higher concentrations, the problem becomes more complex and other phase structures appear. Figure 2.19 shows some of the structures that occur as the surfactant concentration increases. The most recently studied structure is the long branched rod-like structures known as 'worm-like' micelles, for which a large amount of literature has been produced in the past fifteen years. The more organized structures, such as the liquid crystalline mesophases illustrated as (3), (4) and (5) in Figure 2.19, have been known since the early days of soap making when they were referred to as 'middle', 'viscous' and 'neat' phases, respectively. All of these four phases (2–5) are viscoelastic but with different textures. The more fluid ones can be useful for thickened detergent systems, with the lamellar phase giving us bar soaps. Laughlin [16] has produced a particularly useful text describing the phase behaviour and recognition of the various phases. The simplest way to observe

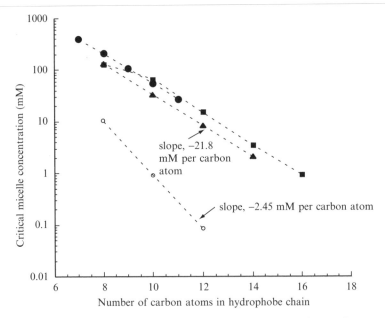

Figure 2.18. The variation of critical micelle concentration with surfactant chain length: ○, CnEO6; ■, CnTAB; ▲, CnSO$_4$Na; ●, CnCOONa (EO, ethylene oxide; TAB, trimethylammonium bromide).

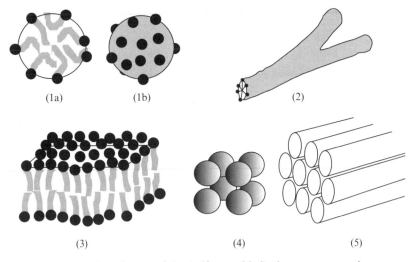

Figure 2.19. Examples of some of the (self-assembled) phase structures that can occur at temperatures above the cmc with increasing concentration of surfactant: (1) spherical micelle ('a' as cross-section); (2) 'worm-like' micelle; (3) lamellar phase; (4) cubic phase; (5) hexagonal phase.

the phase behaviour is to place a small amount of the solid surfactant on a microscope slide in contact with water. This is then viewed with a polarizing microscope. In the contact zone between the solid surfactant and the water, there is a progression of concentration from an isotropic solution through to the solid phase. The refractive index is different in different directions for both the hexagonal and lamellar phases. This results in quite distinct optical patterns for each phase. With the cubic structure, the refractive indices are the same along each of the three axes and no patterns are observed. The sample can also be heated and the changes from one phase to another observed. Usually, the phase behaviour is less sensitive to changes in temperature than it is to changes in concentration, that is, the phase boundaries do not change very much as the temperature is raised, with the major change being the loss of that phase as an isotropic solution is formed. With non-ionic surfactants, the formation of a second, co-existing liquid phase is observed as the *cloud point*. This is the result of the balancing of the hydrophobic effect of the tails and the head group hydration. The latter reduces as the hydrogen-bonding structure lessons with temperature and a second surfactant-rich phase is produced, hence giving the clouding phenomenon. The addition of electrolytes decrease the cloud point while the addition of co-solvents, such as ethanol or propanol, increases the temperature at which we observe the phenomenon. We need to operate close to the cloud point, ca. $\pm 10\,°C$, if we wish to use the surfactant efficiently as a dispersing aid [20]. This maximizes the interfacial responsiveness of the molecules.

Other structures can also be found. Di-chain surfactants, such as lecithin, for example, can form spherical bilayer structures. These may consist of just a single bilayer – rather like a cell – and are known as *vesicles*. Spherical structures made up of multiple bilayers can also be formed. We can summarize by noting that the richness of surfactant phase behaviour is due to the combination of packing constraints and the free energy changes associated with hydrophobe–water interactions, the oil–water interface, and head group interactions, which include charge for ionic surfactants – much easier to study than to model!

8.3 Macromolecular Surfactants

This group of surfactants consists of large, mostly non-ionic molecules with molecular weights in the range $1 \times 10^3 - 3 \times 10^3$ Da. A variety of structures can be produced, with the various types and their uses being given in a review by Hancock [20], which includes, in addition, much information on short-chain non-ionics. Just like short-chain surfactants, the molecules consist of a hydrophobic part and a hydrophilic part. Poly(ethylene oxide) chains are a common choice for the hydrophile as this polymer has good water solubility at molecular weights $> 10^3$ Da. On the other hand, poly(propylene oxide) has poor

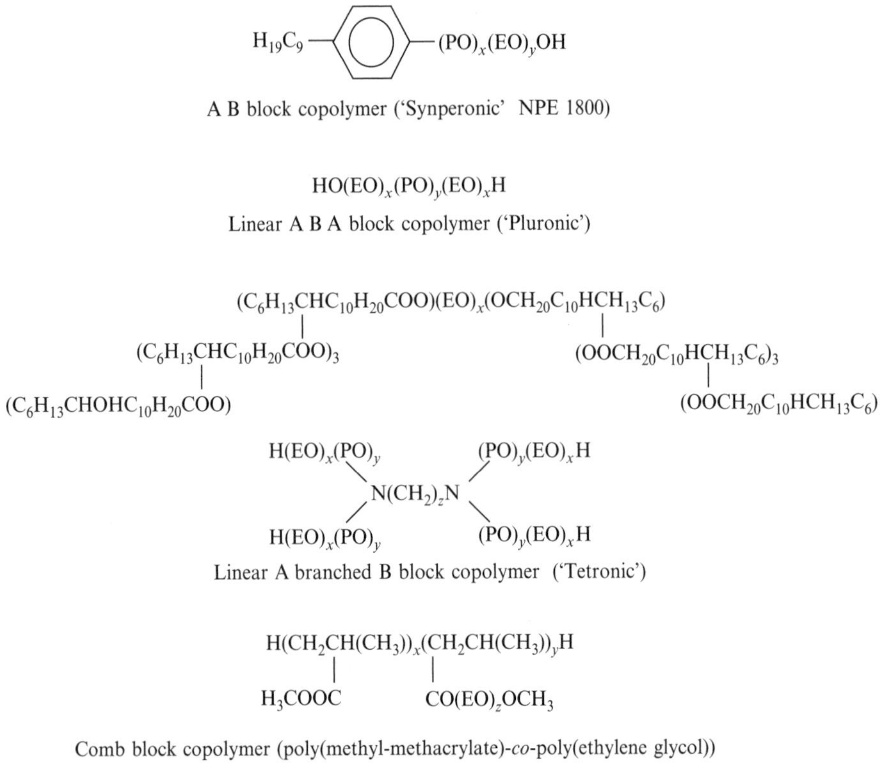

$$H_{19}C_9 - \left\langle \bigcirc \right\rangle - (PO)_x(EO)_yOH$$

A B block copolymer ('Synperonic' NPE 1800)

$$HO(EO)_x(PO)_y(EO)_xH$$

Linear A B A block copolymer ('Pluronic')

$$(C_6H_{13}CHC_{10}H_{20}COO)(EO)_x(OCH_{20}C_{10}HCH_{13}C_6)$$

$$(C_6H_{13}CHC_{10}H_{20}COO)_3 \qquad\qquad (OOCH_{20}C_{10}HCH_{13}C_6)_3$$

$$(C_6H_{13}CHOHC_{10}H_{20}COO) \qquad\qquad (OOCH_{20}C_{10}HCH_{13}C_6)$$

$$
\begin{array}{ccc}
H(EO)_x(PO)_y & & (PO)_y(EO)_xH \\
& \diagdown\diagup \\
& N(CH_2)_zN \\
& \diagup\diagdown \\
H(EO)_x(PO)_y & & (PO)_y(EO)_xH
\end{array}
$$

Linear A branched B block copolymer ('Tetronic')

$$H(CH_2CH(CH_3))_x(CH_2CH(CH_3))_yH$$

$$H_3COOC \qquad\qquad CO(EO)_zOCH_3$$

Comb block copolymer (poly(methyl-methacrylate)-*co*-poly(ethylene glycol))

Figure 2.20. The structures of some macromolecular surfactants [20]: EO, ethylene oxide; PO, propylene oxide.

water solubility at this molecular weight and is therefore a useful choice as a hydrophobe, either alone or in conjunction with other hydrophobic molecules. The structures of some commercially important macromolecular surfactants are shown in Figure 2.20. Such structures can vary from simple linear block structures to sophisticated branched structures resembling 'brushes' or 'combs'.

8.4   Choices of Surfactants for Applications

This is a major problem to most workers in the colloid field, especially when the choice is not limited to those with government approval for the particular product. Hancock [20] gives a useful review of the uses of small and large non-ionic surfactants in a wide variety of applications. However, some general observations are appropriate at this point. Short-chain anionic surfactants are very widely used as stabilizers and wetting agents. These materials are mobile, adsorb to hydrophobic surfaces and form stable films. The sulfates and sulfonates are less vulnerable to the calcium ions in hard water than the potassium salts of fatty acids, where the calcium binds strongly to the carboxyl groups and precipitates the calcium salt as the familiar 'scum'. These materials are widely used as stabilizers and cleaning agents. The cationics interact with biological membranes, which are invariably negatively charged, and are useful as germicides. Coco-betaines and similar ionic materials have relatively low biological activity and can be found in personal-care products such as shampoos. Short-chain non-ionics form more mobile, less rigid films than similar chain length anionics and are useful in low-foaming cleaning systems. The low rigidity of the surfactant films allows the thermal fluctuations to become large enough to aid collapse of the films. Rigid films damp out this motion and are thus more stable. These surfactants are also useful as stabilizers in systems which have too high an electrolyte concentration for charge stabilization to work adequately.

Macromolecular surfactants do not make good wetting agents or emulsifiers. For this purpose, we require small molecules which diffuse rapidly and stabilize new interfaces by adsorption, building charge density and lowering interfacial tensions in the process. However, macromolecular surfactants make excellent stabilizers of emulsions or solid particles. The lower mobilities and large moieties which are insoluble in the continuous phase mean that they are reluctant to leave the interface. Typically, the molecular weights of the soluble and insoluble parts of the molecules are similar. This maximizes the interfacial concentrations of the stabilizing elements. We should also note that these large molecules can show some self-assembly tendencies which can result in multilayer formation at the interfaces, hence leading to even greater stabilities. Poly(12-hydroxystearic acid)-*co*-poly(ethylene glycol) is a good example of this. This surfactant is an excellent stabilizer of high-phase-volume water-in-oil emulsions, such as can be found in cosmetic moisturizing creams. The water droplets can be surrounded by a trilayer of surfactant in some emulsions. In addition, the macromolecular materials are usually less sensitive to temperature or electrolyte levels.

8.5 Proteins at Surfaces

Large protein molecules should also be thought of as macromolecular surfactants, albeit with quite special properties as it is typical of biological systems to have multiple roles. Proteins have a stabilizing function in naturally occurring systems, with milk being a good example in which β-casein species stabilize the fat droplets. Of course, proteins are also widely employed as stabilizers in processed foods. However, it is not just in foodstuffs and pharmaceuticals that they have applications. They have been used as stabilizers for inorganic particles in paints, inks and photographic film, for example.

Adsorption to a hydrophobic surface is by the attachment of the hydrophobic regions and may or may not involve the opening out of the protein structure, i.e. denaturing of the protein, as it adsorbs [21]. Some proteins are relatively unstructured (such as β-casein), while globular proteins have much more structure which is little changed by adsorption. The attachment occurs via a large number of segments, and like synthetic polymers, desorption is a very slow (even unlikely) process, although if the system is challenged by the addition of small mobile surfactants, desorption can be induced. These are mobile enough to adsorb onto surface sites from where a protein segment has temporarily desorbed, thus preventing reattachment. With an adsorption energy of ca. $0.5k_B T$ [21], we should visualize the adsorption at each site as being a dynamic process but, with attachment at several hundred sites, this will not lead to the loss of the whole molecule from the surface unless each site is immediately taken up by a competitor.

Just as adsorption equilibrium is a slow process with synthetic polymers, so it is with proteins. Rapid stirring, of course, limits the diffusion process to the movement across the laminar boundary layer of fluid very close to the surface but the approach to the final conformation and packing density can be a slow process. In the final state, the stabilizing protein film provides a very robust form of stabilization in the form of electrosteric stabilization (see Chapter 3). There is an electrostatic component, although this is sensitive to pH changes as the major component results from the presence of carboxyl groups. This is enhanced by a steric component which resists the local increase in concentration of the stabilizing moieties as two surfaces approach, as well as moving the origin of the electrostatic component away from the surface. Finally, the protein films are viscoelastic and this damps out the thermal interfacial fluctuations, which produce coalescence of emulsions and foams.

Biodegradability is an additional advantage when using proteins as stabilizers, although this may be too rapid, and toxicity is less likely to be a problem. The aliphatic alcohol ethoxylates are viewed as being acceptable in their biodegradability behaviour, unlike the nonylphenol ethoxylates. The toxicity of both to fish and other species though can be a problem in some instances.

REFERENCES

1. L. H. Peebles, *Molecular Weight Distribution in Polymers*, Wiley-Interscience, New York, 1971.
2. P. J. Flory, *Principles of Polymer Chemistry*, Cornell University Press, Ithaca, New York, 1953.
3. P. J. Flory, *The Statistical Mechanics of Chain Molecules*, Wiley-Interscience, New York, 1969.

4. H. Yamakawa, *Modern Theory of Polymer Solutions*, Harper and Row, New York, 1971.
5. P.-G. deGennes, *Scaling Concepts in Polymer Physics*, Cornell University Press, Ithaca, New York, 1979.
6. M. Doi and S. F. Edwards, *The Theory of Polymer Dynamics*, Oxford University Press, Oxford, UK, 1986.
7. P. J. Flory, *Discuss. Faraday Soc.*, **49**, 7 (1970).
8. J. M. H. M. Scheutjens and G. J. Fleer, *J. Phys. Chem.*, **83**, 619 (1979).
9. P.-G. deGennes, *Macromolecules*, **13**, 1069 (1980).
10. P.-G. deGennes, *Adv. Colloid Interface Sci.*, **27**, 189 (1987).
11. M. Bodanecky and J. Kovar, *Viscosity of Polymer Solutions*, Elsevier, Amsterdam, 1982.
12. J. Brandrup and E. H. Immergut (Eds), *The Polymer Handbook*, 2nd Edn, Wiley, New York, 1989.
13. M. Kerker, *The Scattering of Light and Other Electromagnetic Radiation*, Academic Press, New York, 1969.
14. M. B. Huglin (Ed.), *Light Scattering from Polymer Solutions*, Academic Press, New York, 1972.
15. K. Shinoda (Ed.), *Solvent Properties of Surfactant Solutions*, Marcel Dekker, New York, 1967.
16. R. G. Laughlin, *The Aqueous Phase Behaviour of Surfactants*, Academic Press, New York, 1994.
17. D. H. Everett, *Basic Principles of Colloid Science*, The Royal Society of Chemistry, London, 1988.
18. B. Lindman, H. Wennerström and H. F. Eicke (Eds), *Micelles*, Springer-Verlag, Heidelberg, Germany, 1980.
19. P. Mukerjee and K. J. Mysels, 'Critical Micelle Concentrations of Aqueous Surfactant Systems', NSRDA-NBS-36, US Government Printing Office, Washington, DC, USA, 1971.
20. R. I. Hancock, 'Macromolecular Surfactants', in *Surfactants*, T. Th. Tadros (Ed.), Academic Press, London, 1984, pp. 287–321.
21. E. Dickenson, *Introduction to Food Colloids*, Oxford University Press, Oxford, UK, 1992.

Chapter 3

Interactions between Colloidal Particles

1 INTRODUCTION

In Chapter 1, we started our discussion of colloidal systems with the influence of size on the timescale of the motion of the primary components, that is, the particles. In this present chapter, we will extend that discussion to how the particles interact with each other and how these interactions result in the various structures found in colloidal systems. This means that the focus will be on describing the energies of the interactions. Usually, we will be thinking in terms of a potential energy and will not be too concerned about kinetic energy. It is also usual to think in terms of Newtonian mechanics. We can visualize two particles close to each other, say at a centre-to-centre separation of r, and ask how much work would we have to do to separate them to some large distance apart. This is the potential energy of the particle–particle interaction and is termed the *pair potential*, $u(r)$. This is calculated from how much force is required to move the particles. So, if the force at distance r is $f(r)$ then we obtain the pair potential from the following:

$$u(r) = -\int_r^\infty f(r)\mathrm{d}r \qquad (3.1a)$$

and of course:

$$\frac{\mathrm{d}u(r)}{\mathrm{d}r} = -f(r) \qquad (3.1b)$$

Colloids and Interfaces with Surfactants and Polymers – An Introduction   J. W. Goodwin
© 2004 John Wiley & Sons, Ltd   ISBN: 0-470-84142-7 (HB)   ISBN: 0-470-84143-5 (PB)

Therefore, the particles can be thought of as though they are interacting via 'colloidal springs' and it is the nature of these springs that we need to describe. Let us also recall that the spring constant (or modulus) is the rate of change of the force with the distance:

$$\frac{d^2 u(r)}{dr^2} = y(r) \tag{3.2}$$

It is easier to deal with particles in an equilibrium state but we should never forget the timescales as the systems that we use everyday may be far away from their equilibrium state. Formally, the interaction energy is a free energy which should include both the enthalpic and entropic contributions of *all* of the components in the system, whether particles, solvent molecules, small ions, surfactant molecules or polymer molecules. This means that any expression for the interaction energy between any two particles would take into account an average contribution from all of the other components. This is the *potential of mean force*. There are two approaches that are possible to estimate this, i.e. we may determine it from the equilibrium structure of the colloidal system, or we can produce a mathematical model of the system. The former route may be possible in some simple idealized systems, although the latter is very difficult to attain. Thus, what we will do is to try and estimate the various contributions to the potential and then make the assumption that they can be added together to give the total potential. In order to achieve this, we will use relatively simple models which have enough information to get us close to where we wish to go and give an adequate description of what we observe. It is important to remember that these are simple models and we should not be too surprised if they only agree with experimental results under limited conditions. Often though, they will adequately serve our purpose.

There is an excellent publication by Isrealachvili [1] which gives a detailed outline of both intermolecular forces and the interactions between particles. As this present work is an introductory text, we shall be just working with the salient features and if the interested reader should need more details, then the cited text will supply it. A number of contributions to the interactions between particles can be identified. Some of the interparticle forces work to bring particles into close contact, while others act to separate them. As we shall see, it is the interplay between these that results in the final state of our dispersions. If we understand the origins of these forces, it becomes possible to modify them by control of the chemical environment or the chemical architecture of the components of the dispersion. Each of the major contributions to the net or total interparticle force can be described by using physical models. These can become quite complex but they are also centred around quite simple initial models. This makes it easy to understand the underlying concepts even if the full manipulation becomes difficult, but then the application to complicated systems that we wish to use can appear a daunting prospect. However, the same general principles apply and, although an accurate predictive calculation may not be within our reach for a particular product, the choice of experiments will be much better focused.

2 INTERMOLECULAR ATTRACTION

The starting point for our discussion must be an examination of the forces that occur between molecules. Some of these interactions are strong, and hence are long-lived, such as the covalent bond, or the weaker, more transient hydrogen bond. The origin of the attraction between particles does not lie in these interactions but in the weaker interactions that are often referred to as van der Waals interactions. These are the forces that account for the non-ideality of gases and account for the deviations from the simple behaviour described by the ideal gas equation:

$$\frac{PV}{T} = NR = nk_B \tag{3.3}$$

where the pressure is P, the volume V and the temperature T; n is the number of gas molecules in the volume, or N if we use the number of moles, and k_B is the Boltzmann constant. There are several interactions that can occur which are electrodynamic in origin and the traditional description separates these into three distinct forms. We will discuss the origins of these but only at a level sufficient to enable us to see the underlying mechanism for the attraction.

2.1 Keesom Interaction

It is not much of a surprise to find that the molecules in a gas such as hydrogen chloride have a strong permanent dipole moment due to the polarization of the covalent bond. The dipoles tend to align and this will be the preferred arrangement. This dipole–dipole attraction is known as the *Keesom interaction* and we can write the interaction free energy at an intermolecular distance r as follows:

$$u(r)_K = -\frac{C_K}{r^6} \tag{3.4}$$

where C_K is a constant which depends on the particular type of molecule being considered. For example, if we have two identical molecules of dipole moment μ:

$$C_K \propto \mu^4 \tag{3.5}$$

Because of the marked dipole alignment, the rotational motion of the molecules is restricted and we should be thinking of this as a 'long-time' interaction, that is, a low-frequency interaction.

2.2 Debye Interaction

This is the type of interaction that occurs between a polar molecule and a non-polar one. The dipole on the polar molecule polarizes the electron cloud of the non-polar molecule. The molecular rotation is still occurring and thus we could think here of the frequencies associated with the interaction as those in the microwave region. The interaction free energy for this dipole-induced dipole interaction can be described by a similar expression to that used for the dipole–dipole interaction (Equation (3.4)). In this case, the polarizability of the non-polar molecule is a key feature. It should also be recognized that even for two polar molecules with different dipole moments the net dipoles will be affected by interaction with the adjacent molecules. The following equation illustrates this for molecules '1' and '2':

$$u(r) = -\frac{C_D}{r^6} \tag{3.6}$$

and we note that the interaction constant, C_D, for a system consisting of two different molecules has the following dependence:

$$C_D \propto \left(\alpha_2 \mu_1^2 + \alpha_1 \mu_2^2 \right) \tag{3.7}$$

We are using an example with two permanent dipoles which will orient to maximize the interaction and the timescales will again be long.

2.3 London or Dispersion Interaction

This describes the interaction that results in attraction between non-polar molecules. It is due to the movement of the electron cloud around the atomic nucleus resulting in a dipole that fluctuates. When two atoms come into close proximity, the temporary dipoles become aligned, that is, the fluctuations become coupled, and this is a preferred (or lower) energy state. Such a situation is illustrated in Figure 3.1.

The range of the interaction is similar to the two discussed previously but the timescale of the fluctuations is now that of the electronic transitions and so we should think towards the visible/ultraviolet part of the electromagnetic spectrum. The interaction energy can be written as:

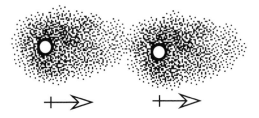

Figure 3.1. The coupling of oscillating atomic dipoles which forms the basis of the London (or dispersion) force between atoms.

$$u(r) = -\frac{C_L}{r^6} \qquad (3.8)$$

Now, the London constant, C_L, for two similar atoms is proportional to the ionization energy of the outer electrons, $h\nu_I$, where h is the Planck constant, and the polarizability, α, as follows:

$$C_L \propto h\nu_I \alpha^2 \qquad (3.9)$$

and for two different types of molecule:

$$C_L \propto h\left(\frac{\nu_{I1}\nu_{I2}}{\nu_{I1} + \nu_{I2}}\right)\alpha_1\alpha_2 \qquad (3.10)$$

It turns out that this is a particularly important type of intermolecular interaction. This is because it is much larger than the Keesom or Debye contributions in nearly all cases of colloidal materials. Water is a notable exception, with the dispersion interaction contributing only a quarter of the total. An important feature of this interaction is that there is only a weak tendency to change the orientation of neighbouring molecules. In the solid phase, this is important as the fluctuations can still couple without requiring motion of the molecules to maximize their alignment, as would be the case for molecules with permanent dipoles.

Let us now consider the implications of this 'attractive' interaction with the normal alkane series. These have particular relevance to many colloidal systems as many of our surfactant systems have a linear aliphatic chain as one element of their composition. This element has a tendency to self-assemble at interfaces where the chains come into close proximity. The cohesive energy of the solid hydrocarbon is estimated from the measured latent heats of melting and vaporization. Here, we will use the data given by Isrealachvili [1]

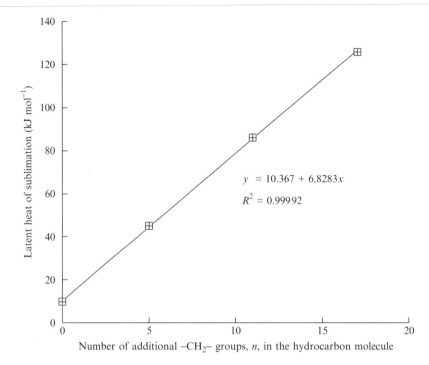

Figure 3.2. The latent heat of sublimation for a series of linear alkanes, $HCH_2-(CH_2)_n-H$ [1].

and use his simple model calculation. Figure 3.2[1] shows the value of the molar cohesive energy as $-CH_2-$ groups are added to the chain. The slope gives the value per additional $-CH_2-$ as $6.8\,kJ\,mol^{-1}$. This is equal to $\sim 2.75k_BT$ at 25 °C, showing that these non-polar molecules are quite 'sticky'. Taking each chain as being surrounded by six nearest neighbours, and summing the interactions over the neighbouring groups up and down the chain, as well as around it, gave a value of $\sim 6.9\,kJ\,mol^{-1}$.

This simple result is a very important illustration of how the dispersion forces can apparently be additive to a good approximation. This will lead us to consider the behaviour of the relatively large groups of molecules that make up colloidal particles. However good this approximation appears to be in some simple cases, it should be remembered that the interaction of each

---

[1]This figure shows the mathematical relationship for the straight line drawn through the experimental points. R is known as the *correlation coefficient*, and provides a measure of the 'quality of the fit'. In fact, R^2 (the *coefficient of determination*) is used because it is more sensitive to changes. This varies between -1 and $+1$, with values very close to -1 and $+1$ pointing to a very tight 'fit' of the data.

molecule is affected by all of the neighbouring fields and so that pair-wise additivity is only an approximation.

2.4 A More Generalized Approach

All of the three interactions described above are based on the attraction between dipoles and have the same distance-dependence. Therefore, we can expect that we could handle these in one general description. The London constant shown in Equations (3.9) and (3.10) is written as depending on the ionization of the outer electrons. Other electronic transitions also take place so that contributions at other frequencies can also occur. Thus, we must think in terms of the full range of frequencies ranging from those of a few hertz up to the ultraviolet region at $\sim 10^{16}$ Hz. Furthermore, we have only considered pairs of atoms or molecules interacting in the absence of any intervening medium. The *dielectric constant*, or *relative permittivity*, of the medium is the important factor here. If a sphere is placed in a medium of the same permittivity and an external field applied to the system, the sphere will not be polarized. If it has a dielectric constant which is either larger or smaller than the medium, then it will be polarized. This effect is the basis for electro-rheological fluids whose solid/liquid properties are readily and rapidly controlled by the application of an electric field. In order to achieve the high fields necessary for strong effects, the choice is to use a low dielectric constant for the medium with a higher value for the material making up the particles. It is the difference in dielectric constant which is important and we should think in terms of an 'excess polarizability' [2]. It is important to recognize that for the case of permanent dipoles there is an alignment contribution to the polarizability as well as the electronic component. It is only the latter, of course, that we need to consider with non-polar molecules but the full spectral range would need to be included for full prediction. The dielectric behaviour as a function of frequency to give $\varepsilon(\nu)$ is a tractable experiment at lower frequencies and the refractive index, $n(\nu)$, is a viable measure of the dielectric behaviour at the higher end of the spectrum (recall that $\varepsilon(\nu) = n^2(\nu)$).

The general description of the interaction was given by McLachlan [3] as a summation over the range of interaction frequencies for molecules '1' and '2', interacting in a medium '3', as follows:

$$u(r) = -\left[\frac{3k_B T \alpha_1(0)\alpha_2(0)}{(4\pi\varepsilon_3(0)\varepsilon_0)^2 r^6} + \frac{6k_B T}{(4\pi\varepsilon_0)^2 r^6}\sum_{n=1}^{\infty}\frac{\alpha_1(i\nu)\alpha_2(i\nu)}{\varepsilon_3^2(i\nu)}\right] \tag{3.11}$$

This summation is carried out over all frequencies so that all of the interactions are captured. The first term on the right-hand side of the above equation is the 'zero-frequency' term. Values for the latter are what we will usually find in data tables. The second term on the right-hand side of the equation contains the contributions for all other frequencies, with the steps being in terms of $k_B T$:

$$n = \left(\frac{\nu_n h}{2\pi}\right) / k_B T \qquad (3.12)$$

However, this term uses the frequency dependence of the polarizability at complex frequencies, $i\nu$. Here, i indicates the *quadrature* or *imaginary* component of the frequency. This is the dissipative component as we are looking at the interaction of the oscillating electromagnetic field with the molecules and not the transmitted component of the field.

3 COMPLEX NUMBER MANIPULATION

We need to use complex number notation whenever we wish to describe the behaviour of oscillating fields. This is something that we have to do frequently in physical science. For example, we will use this in the later chapter in this volume which deals with rheology when, for example, we describe what happens when we apply an oscillating force mechanically to a concentrated colloidal system. Here, we use the algebra to separate the elastic storage of work done from that dissipated by viscous flow. (The response is similar to that we would get from oscillating a dampened spring such as that used in an automotive suspension unit.) Another common example is that of an AC electric circuit with a capacitor and a resistor in series. At zero frequency (DC voltage), we can store electrical energy in the capacitor and we would have little interest in the resistor. As we increase the frequency of the AC voltage, we store less in the capacitor and dissipate energy in the resistor. Let us come now to examples closer to the subject of this chapter and consider electromagnetic radiation. In the usual science courses, we learn about spectra and how the applied radiation interacts with matter. We are familiar with the fact that light can pass through a solution but we may find the intensity reduced at some particular frequencies, perhaps in the UV, the IR and the microwave regions. However, at the same time we talk of a refractive index. What we mean here is what we would measure by transmission at some frequency. It is, in fact, a complex number in which we should include dissipative terms as well as storage, just like the AC circuit. The same applies to dielectric constants. We are referring to the static (low-frequency plateau) value. We can carry out the measurements at say 10^3 Hz and at higher frequencies we find interesting behaviour as polar molecules respond and give dissipative contributions as the timescales become too short for their motion.

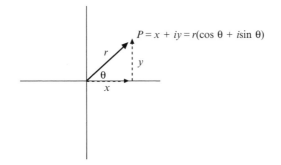

Figure 3.3. Illustration of complex number notation.

Initially, the use of complex number algebra appears much more difficult to many scientific workers than is actually the case. The biggest barrier is the nomenclature as when we say the number $P = x + iy$, where $i^2 = \sqrt{-1}$ and that x is the *real* part of P and y is the *imaginary* part! All we have to remember is that the ordinary rules of algebraic manipulation apply and that whenever we arrive at i^2 we write -1 and also remember that $1/i = -i$. Even this looks a little intimidating without some practice. It is useful just to think of the notation as a simple way of writing down how we are dividing P, the numerical value that we measure, into two contributions, x and y. Figure 3.3 shows this in graphical form, where P is given in both Cartesian coordinates and polar coordinates.

To illustrate how this works, let us consider the polarizability, which is a function of frequency. We can write the total polarizability as the sum of any dipole (permanent or induced) alignment term, $\alpha_a(\nu)$, and the electronic polarizability of a molecule, $\alpha_e(\nu)$ which is responsible for the dispersion force [1, 4], as follows:

$$\alpha(\nu) = \alpha_a(\nu) + \alpha_e(\nu) \tag{3.13a}$$

This may be written in more detail as:

$$\alpha(\nu) = \frac{\mu^2}{3k_B T(1 - i\nu/\nu_{\text{rotation}})} + \frac{\alpha_e}{\left[1 - (\nu/\nu_I)^2\right]} \tag{3.13b}$$

where ν_{rotation} is the rotational frequency of the dipolar molecule and ν_I is the ionization frequency (so here we are assuming that there is only one absorption frequency, ν_I).

The quadrature (imaginary) component is:

$$\alpha(i\nu) = \frac{\mu^2}{3k_B T(1 - i^2\nu/\nu_{\text{rotation}})} + \frac{\alpha_e}{\left[1 - (i\nu/\nu_I)^2\right]} \tag{3.13c}$$

while for the range of n frequencies we obtain:

$$\alpha(i\nu_n) = \frac{\mu^2}{3k_B T(1 + \nu_n/\nu_{\text{rotation}})} + \frac{\alpha_e}{\left[(1 + (\nu_n/\nu_I)^2)\right]} \tag{3.13d}$$

Therefore, the utilization of Equation (3.11) should appear less daunting as the frequencies are just those where we observe adsorption. Consider now the situation where we have two similar non-polar molecules interacting in a gas – methane would be a good example. Equation (3.11) becomes ($\varepsilon_3 = 1$ as the interaction is across a vacuum):

$$u(r) = -\left[\frac{3k_B T\alpha_{\text{CH}_4}^2(0)}{(4\pi\varepsilon_0)^2 r^6} + \frac{6k_B T}{(4\pi\varepsilon_0)^2 r^6}\sum_{n=1}^{\infty}\alpha_{\text{CH}_4}^2(i\nu)\right] \tag{3.14}$$

We can replace the summation by an integration. Using Equation (3.12), we have:

$$dn = (h/2\pi k_B T)d\nu \tag{3.15}$$

and so:

$$u(r) = -\left[\frac{3k_B T\alpha_{\text{CH}_4}^2(0)}{(4\pi\varepsilon_0)^2} + \frac{3h}{(4\pi\varepsilon_0)^2\pi}\int_{\nu_1}^{\infty}\alpha_{\text{CH}_4}^2(i\nu)d\nu\right]\frac{1}{r^6} \tag{3.16}$$

In Equation (3.16), when frequencies approach the visible range, the second term on the right-hand side becomes dominant. At frequencies below the microwave region, only the first term on the right-hand side is important and this is, of course, a constant. Methane has no permanent dipole and so the frequency-dependent polarizability from Equation (3.13d) is:

$$\alpha_{\text{CH}_4}(i\nu) = \frac{\alpha_{e\text{CH}_4}}{1 + (\nu/\nu_I)^2} \tag{3.17}$$

Substituting this into Equation (3.16) and integrating yields the London dispersion result as we are just using the ionization frequency:

$$u(r) \approx -\left[\frac{3h}{(4\pi\varepsilon_0)^2}\alpha_{e\text{CH}_4}^2\nu_I\right]\frac{1}{r^6} \tag{3.18}$$

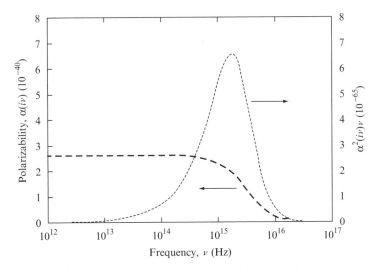

Figure 3.4. Variation of the frequency-dependence of the quadrature component of the polarizability of methane.

Hence, we may write the interaction energy with the London constant as follows:

$$u(r) = -\frac{C_L}{r^6} \tag{3.19}$$

The electronic polarizability for methane is $2.89 \times 10^{-40}\,\text{C m}^2\,\text{J}^{-1}$ and the ionization frequency is $3.05 \times 10^{15}\,\text{s}^{-1}$, which gives a value of $C_L = 1.02 \times 10^{-77}\,\text{J m}^6$.

A plot of $\alpha^2_{CH_4}(iv)v$ as a function of v is given in Figure 3.4 and this shows the spectral region which dominates the attraction between the methane molecules. Note how it is centred in the ultraviolet where we are accustomed to seeing electronic excitations.

4 DISPERSION FORCES BETWEEN PARTICLES

We now have some understanding of how atoms and molecules attract each other. In addition, we have indicated that considering the dispersion forces to be additive can be a reasonable approximation as we start to work with larger molecules. This clearly has importance when we are going to consider surfactants and polymers, but it may seem surprising that we can extend this approximation to describing the interaction between two particles. This

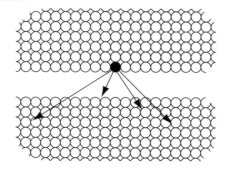

Figure 3.5. The summation of the atomic dipolar interactions between two slabs of material.

approach laid the foundation for our understanding of why particles coagulate and was pioneered by workers such as de Boer[5], Hamaker[6], Derjaguin[7] and Langbein[8] using London's approximation based on the interaction at a frequency around the ionization frequency. This is where we will begin.

The starting point is to consider the interaction between a single molecule and a slab of material and then extend this to two slabs interacting. As we are assuming that the energies are additive, we simply use Equation (3.19) and add up the interactions between our reference molecule and all of the molecules in the slab of material (illustrated in Figure 3.5). Therefore:

interaction energy = sum for molecules with each in lower slab

Next, we add this up for all of the molecules in the upper slab and so we now have:

interaction energy = sum of all molecules in upper slab

$\times$ (sum for molecules with each in lower slab)

The problem is that in order to calculate r for each interaction we would need to know the detailed structure. It is simpler to use the number density of molecules in the slab, ρ_N, and to integrate over the volumes. This 'semi-continuum' approach is in the spirit of the additivity assumption and would only become a problem at very close separations. This results in the energy as follows:

interaction energy = $C_L \rho_N^2$ (a geometric term)

and we have a general approach which has the characteristic of the material, i.e. the electronic polarizability, the ionization frequency, and the square of the product of the density and the molar mass as a front factor to a term based on the shape of the material and the separation distance between the two bodies. Depending on the shape of the two particles, solving the integrals can be non-trivial but it is essentially a problem of geometry and calculus and not one of physics. Examples of the algebraic manipulations are given in the literature [1, 9, 10], although the purpose of this present text is just to illustrate the underlying mechanisms of colloid science. So, for example, for two slabs which we can consider to be infinitely thick, the energy of unit area of one slab interacting with the whole of the other slab is:

$$V_A = -\frac{\pi C_L \rho_N^2}{12 H^2} \tag{3.20}$$

Here, V_A is used to represent the dispersion energy between two slabs a distance H apart and is the energy per unit area of surface. The numerator is the material property and the denominator arises from the geometry. It is common to express the material properties as a single material constant, i.e. the *Hamaker constant*:

$$V_A = -\frac{A_{11}}{12\pi H^2}, \text{ where } A_{11} = \pi^2 C_L \rho_{N1}^2 \tag{3.21}$$

The subscript '11' indicates that the interaction is between two slabs of the same material. So, if we had a slab of material '1' interacting with material '2', the Hamaker constant notation would indicate this, namely:

$$A_{12} = \pi^2 C_{L12} \rho_{N1} \rho_{N2} \tag{3.22}$$

If the 'semi-infinite' slabs are replaced by plates of finite thickness, t, Equation (3.21) has a slightly more complicated form:

$$V_A = -\frac{A_{11}}{12\pi} \left[\frac{1}{H^2} + \frac{1}{(H+2t)^2} - \frac{2}{(H+t)^2} \right] \tag{3.23}$$

Equation (3.23) was used to calculate the attractive energy for two small plates as a function of thickness, using a value of A_{11} of 10^{-19} J and plate areas of $0.1\,\mu m^2$. These values would be similar to those found for clay particles. It is clear from Figure 3.6 that the finite thickness of the plate is only important for thin plates and also that the variation of the interaction with distance is greater for thin plates.

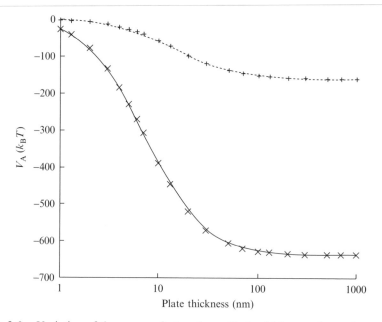

Figure 3.6. Variation of the energy of attraction with the thickness of two plates, using Equation (3.23): ×, thickness = 10 nm; +, thickness = 20 nm.

However, we still have the energy of attraction equal to the product of a material constant and a geometric term which has been derived from an integration over the particle volumes. This is the case whatever the shape of the particles. For example, if we have two identical spheres of radius a and centre-to-centre distance r, then when the spheres are close together the attractive energy is:

$$V_A(r) = -\frac{A_{11}a}{12(r - 2a)} \tag{3.24a}$$

or:

$$V_A(H) = -\frac{A_{11}a}{12H}, \text{with } H = (r - 2a) \text{ and } H \ll a \tag{3.24b}$$

Here, H is the closest distance between the surfaces of the spheres. This represents the total energy of interaction and not an energy per unit area of surface. The more general expression for two spheres of radii a_1 and a_2 is [6]:

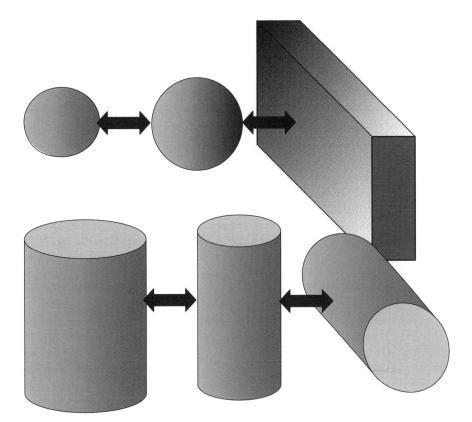

Figure 3.7. Illustrations of interactions between different geometric forms: sphere–sphere; sphere–plate; parallel cylinder; crossed cylinder. Note that the latter is equivalent to the sphere–plate interaction if both cylinders have the same radius.

$$V_A(r) = -\frac{A_{11}}{6}\left\{\frac{2a_1a_2}{r^2-(a_1+a_2)^2} + \frac{2a_1a_2}{r^2-(a_1-a_2)^2} + \ln\left[\frac{r^2-(a_1+a_2)^2}{r^2-(a_1-a_2)^2}\right]\right\} \quad (3.25a)$$

and if the spheres are the same radii:

$$V_A(H) = -\frac{A_{11}}{12H}\left[1 + \frac{H}{2a+H} + \frac{H}{a}\ln\left(\frac{H}{2a+H}\right)\right] \quad (3.25b)$$

The problem in calculating the dispersion interaction is the determination of the geometric term which can become quite complex for relatively simple geometric forms (see Figure 3.7). Some of these, such as parallel cylinders, crossed cylinders and a sphere and a plate (the latter two result in the same geometric factor), are given in the literature [1, 9–11]. For example, when a spherical particle is close to a plate we have:

$$V_A(H) = -\frac{A_{12}a}{6H} \qquad (3.26)$$

A moment's inspection of Equations (3.24b) and (3.26) shows that the inter-action of a sphere with a thick plate has a geometric factor which is twice that of two similar-sized spheres.

It should be noted that in Equation (3.26) we are considering the material of the sphere to be different from that of the plate – hence the notation A_{12}. This is an appropriate point to include a consideration of the additional problem of when our particles are immersed in a continuous medium and what effect that this will have when compared to the case of two particles *in vacuo*. The effect of the particles being immersed in a medium is to reduce the net interaction. If, for example, the Hamaker constant of the medium approaches that of the particles, the attractive interaction approaches zero. We must think in terms of a *net* or *effective* Hamaker constant. There are approximate estimates that we can use to estimate the effective Hamaker constant (for example, see Hunter [9]). If we have particles of types '1' and '2' immersed in a medium '3' we can write the effective Hamaker constant as the sum of the particle–particle and medium–medium interactions for identical volumes across a vacuum, minus the cross terms:

$$A_{132} = A_{12} + A_{33} - A_{13} - A_{23} \qquad (3.27a)$$

If the two particles are of the same material, this, of course, becomes:

$$A_{13} = A_{11} + A_{33} - 2A_{13} \qquad (3.27b)$$

Now, if we make the assumption that:

$$A_{jk}^2 = A_{jj}A_{kk} \qquad (3.27)$$

then we may write Equations (3.27a) and (3.27b) as follows:

$$A_{132} \approx \left(A_{11}^{\frac{1}{2}} - A_{33}^{\frac{1}{2}}\right)\left(A_{22}^{\frac{1}{2}} - A_{33}^{\frac{1}{2}}\right) \qquad (3.28a)$$

and:

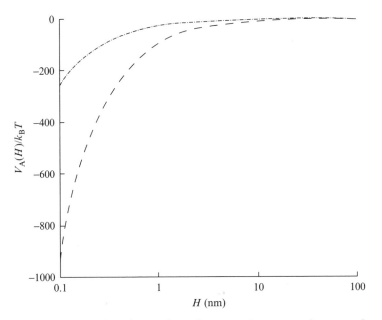

Figure 3.8. The attractive interactions between $2\,\mu m$ tetradecane spheres in water, both with each other and with the polystyrene container wall: (—·—·—) $C_{14}H_{30}$ sphere–sphere; (----) $C_{14}H_{30}$ sphere–wall: $A_{C14} = 5.2 \times 10^{-20}$; $A_{H_2O} = 3.7 \times 10^{-20}$; $A_{PS} = 6.6 \times 10^{-20}$.

$$A_{13} \approx \left(A_{11}^{\frac{1}{2}} - A_{33}^{\frac{1}{2}}\right)^{2} \tag{3.28b}$$

As an example of how these equations are relatively simple to use, let us consider the situation where we have oil droplets (say n-tetradecane), which are $2\,\mu m$ in diameter, dispersed in water with the dispersion in a polystyrene (ps) container. Figure 3.8 shows the interactions between the individual droplets and between the droplets and the container wall at close approach. In this figure, the interactions were calculated by using Equations (3.24) and (3.26) with the effective Hamaker constants from Equations (3.28) for the cases of sphere– sphere and sphere–plate interactions. This figure illustrates that first there is a strong attraction in both cases, even at distances of tens of nanometres, and secondly that the stronger attraction is between the oil droplets and the wall of the container. This is partially because the sphere–plate interaction is twice that of two similar spherical particles, but also because the net Hamaker constant for the interaction is larger. Although both are simple hydrocarbons, polystyrene is a denser molecule with a large amount of aromatic character giving a different electronic polarizability.

5   RETARDED DISPERSION FORCES

The range of the separation axis shown in Figure 3.8 is limited for several reasons. At close separations, the detailed surface structure becomes a limiting factor. We have assumed an atomically smooth surface. This can only be the case for a limited number of conditions such as fluid surfaces, or solids such as carefully cleaved mica. Even with these surfaces there is a limit as the surfaces approach. This occurs as the electron orbital overlap becomes significant. At this point, the free energy becomes dominated by a very short-range repulsion. This is known as *Born repulsion* and is found as the inverse r^{12} dependence in the well-known Lennard–Jones–Devonshire equation [12]. The former limitation though is of even greater importance in most practical systems as we have often intentionally adsorbed material to mitigate the interaction and the outer periphery of any adsorbed layers will not be as well defined as the surface of the adsorbing particle. (The properties of the layer should also be taken into account in estimating the attraction between the surfaces.) The upper limit in Figure 3.8 is partially due to the restrictions for the use of the simplified forms of Equations (3.24) instead of the full Hamaker expressions, Equations (3.25). However, there is another feature that we must discuss at this point. As the distance between the particles increases, the correlation of the oscillating atomic dipoles becomes poorer and the free energy decreases at a greater rate than we would have predicted. This is due to the timescale of the field propagation when compared to the timescale of the oscillation. This is termed *retardation* of the dispersion interaction.

The field propagation time is equal to the separation distance between our surfaces divided by the velocity of light in the intervening material. So, for example, if we consider two surfaces separated by a distance of 30 nm, the propagation time from one surface to the other and then back would be $\sim 10^{-16}$ s. Now, the frequency of the radiation where the strongest interactions are occurring is usually in the near-UV region and so the propagation time represents the time taken for a significant part of an oscillation to occur. In other words, the oscillations will no longer be quite 'in-phase' (i.e. the dipoles are no longer completely aligned) and the attraction is weakened. This weakening will become most marked for the higher-frequency contributions. So, we can conclude that we may have a good approximate description of the attraction between particles at close separation but at separations > 30 nm the values calculated will overestimate the interaction. The indication is also that there should be a frequency-dependence of this effect and a different approach might be sought.

6 THE GENERAL OR LIFSHITZ THEORY OF DISPERSION FORCES BETWEEN PARTICLES

There is an additional problem with particles that are dispersed in a medium such as an electrolyte solution. This is that the low-frequency contributions (i.e. in the extreme case, the interactions of permanent dipoles) are heavily screened by the ionic environment and the low-frequency interaction falls off more rapidly with distance than would be the case in a vacuum. This indicates that we can expect to be running into difficulties with the Hamaker model and should see if the more general approach that was introduced above will help us.

The starting point is to consider two slabs of material, '1' and '2', interacting across a medium '3'. We need to think of each of these materials as slabs of a dielectric material with static dielectric constants of $\varepsilon_1(0)$, $\varepsilon_2(0)$ and $\varepsilon_3(0)$. The interaction energy is a material property (the Hamaker constant) multiplied by a geometric factor. Equation (3.22) defined the Hamaker constant for us and so for the two slabs acting across medium 3 we have:

$$A_{132} = \pi^2 C_{L132} \rho_{N1} \rho_{N2} \tag{3.29}$$

The London term, C_{L132}, is given by Equation (3.11) (McLachlan's equation) which includes the polarizabilities of the molecules making up 1 and 2. The bulk polarizability of slab 1 in the medium 3 can be written in terms of the molecular properties of the material in 1 and the measurable macroscopic dielectric properties, as follows:

$$\rho_{N1} \alpha_1(iv) = 2\varepsilon_0 \varepsilon_3(iv) \frac{\varepsilon_1(iv) - \varepsilon_3(iv)}{\varepsilon_1(iv) + \varepsilon_3(iv)} \tag{3.30}$$

The frequency-dependent Hamaker function is then [1]:

$$A_{132} \approx \frac{3k_{\mathrm{B}}T}{4} \left(\frac{\varepsilon_1(0) - \varepsilon_3(0)}{\varepsilon_1(0) + \varepsilon_3(0)} \right) \left(\frac{\varepsilon_2(0) - \varepsilon_3(0)}{\varepsilon_2(0) + \varepsilon_3(0)} \right)$$
$$+ \frac{3h}{4\pi} \int_{v_1}^{\infty} \left(\frac{\varepsilon_1(iv) - \varepsilon_3(iv)}{\varepsilon_1(iv) + \varepsilon_3(iv)} \right) \left(\frac{\varepsilon_2(iv) - \varepsilon_3(iv)}{\varepsilon_2(iv) + \varepsilon_3(iv)} \right) dv \tag{3.31}$$

The above equation indicates that we should know the dielectric properties over the whole spectral range. There are some interesting implications of Equation (3.31). Let us consider a small particle, '1', dispersed in medium '3', interacting with a larger particle, '2'. If, for example, $\varepsilon_1(iv) < \varepsilon_3(iv) < \varepsilon_2(iv)$, then A_{132} would be negative. This means that a particle of material 1 would be rejected from the medium 3. If the particle of material 2 were liquid, particle 1 could be engulfed. (Engulfment is an important process in biological systems, for example, it is part of our body's defence mechanism where white cells engulf and remove foreign cells such as bacteria in a process known as *phagacytosis*.) After engulfment, particle 1 is in medium 2 and interacting with the surface of 3. In this case, $\varepsilon_3(iv)$ is interchanged with the $\varepsilon_2(iv)$ in Equation (3.31). Both terms in the brackets are negative and A_{123} is positive.

Calculation of the Hamaker 'constant' requires extensive measurements from dielectric spectroscopy. The task is clearly a non-trivial one and the full dielectric data are only available for a very few systems. Polystyrene particles dispersed in aqueous media are a popular model colloidal system and Parsegian and Weiss [13] have carried out the full calculation for this system and compared the results with the Hamaker treatment. Good agreement was obtained but this does not mean that the more complicated Lifshitz analysis should be neglected. It naturally includes retardation effects as well as changes to the intervening media between particles.

Higher-frequency dielectric information is found from the frequency dependent-refractive index data [1, 11]:

$$\varepsilon(iv) = 1 + \frac{(n^2(0) - 1)}{(1 + v/v_e)} \tag{3.32}$$

where v_e is the frequency of the dominant adsorption in the UV and $n(0)$ is the low-frequency value of the refractive index in the visible range. For the dispersion interaction, we can neglect interactions in the microwave region and below. Some approximations for Equation (3.31) are available (e.g. Isrealachvilli [1] and Russel *et al.* [10]). For example, if we have two particles of similar material, '1', interacting in a medium '3', the effective Hamaker 'constant', that is, the distance-dependent value of A which we will write as $A_{131}(H)$, is:

$$A_{131}(H) = \frac{3k_BT}{4}\left(\frac{\varepsilon_1(0) - \varepsilon_3(0)}{\varepsilon_1(0) + \varepsilon_3(0)}\right)^2 + \frac{3hv_e}{16\sqrt{2}}\frac{\left(n_1^2(0) - n_2^2(0)\right)^2}{\left(n_1^2(0) + n_2^2(0)\right)^{\frac{3}{2}}}F(H/\lambda_e) \tag{3.33}$$

Here, $F(H/\lambda_e)$ is a function of the distance between the surfaces relative to the wavelength at the UV adsorption peak. The function is equal to 1 at small distances of separation and reduces quite rapidly with distance H to a value:

$$\frac{H}{\lambda_e} \to \infty; \; F(H/\lambda_e) \sim \left[\frac{2^{\frac{5}{2}}}{\pi n_3(0)\left(n_1^2(0) + n_3^2(0)\right)^{\frac{1}{2}}}\right]\frac{H}{\lambda_e} \tag{3.34}$$

and there is an interpolation formula available [10] which can be used to calculate intermediate values:

$$F(H/\lambda_e) \approx \left\{ 1 + \left[\frac{H}{\lambda_e} \frac{\pi n_3(0) \left(n_1^2(0) + n_3^2(0) \right)^{\frac{1}{2}}}{2^{\frac{5}{2}}} \right]^{\frac{3}{2}} \right\}^{-\frac{2}{3}}$$ (3.35)

So, our computational procedure for estimating the interaction between two spherical particles of the same material dispersed in a medium would be to use Equation (3.25) with an effective Hamaker constant estimated from Equation (3.33) with Equation (3.35) at each separation distance. Although the process appears a little complicated, it is relatively easy to program a computer to carry this out. However, before we can do this we need to take another look at Equation (3.33). The first term in this equation uses the static dielectric constants. This means that we are looking at the contributions from permanent dipoles. Now, in media that are electrolytes, whether made up of aqueous or organic components, the dipole interactions become heavily screened and the first term approaches zero.

7 SUMMARY AND CALCULATION GUIDE

So far in this chapter we have seen how the van der Waals forces acting between molecules arise from the dipolar behaviour of the molecules even when there are no permanent dipoles. The latter are the London dispersion forces. This early treatment assumed that the interaction occurred at a single frequency. The forces may be summed to account for the interaction between particles, as was carried out by Hamaker and de Boer. The interaction was shown to be the product of a material property, a geometric term for the shape of the bodies, and a distance-dependence.

More recently, the analysis has been carried out by treating the particles as macroscopic dielectric materials. This is known as the Lifshitz treatment and has some strong advantages, as follows:

(a) The pair-wise summation is no longer carried out and the macroscopic behaviour automatically includes multi-molecular interactions.
(b) The interaction over the whole spectrum is included so that all of the dipolar interactions are inherent in the computation.
(c) The retardation of the interaction that shows as a reduction in the inter-action free energy as the separation between the bodies is increased can be included as an integral part of the calculation. Moreover, the retard-ation naturally is shown to reduce the high-frequency part of the spectral response.

(d) Any intervening medium is included naturally in the calculation and it is no longer necessary to include combining relationships to provide an approximation to the Hamaker constant.

(e) When the intervening medium is an electrolyte, the low-frequency inter-actions (Keesom and Debye terms) are screened out.

(f) The equations indicate that two bodies of the same material will always attract each other, but also that it is possible to have a repulsive inter-action with the right combination of dielectric properties.

8   CALCULATION STRATEGY

In principle, the full Lifshitz calculation requires a detailed knowledge of the full electromagnetic spectral response and these data are not often available in sufficient detail. However, the approximations given above will usually be sufficient. There are several decisions that we have to make, based on the following questions, listed as follows:

(a) For what purpose are we going to use the results?

(b) Do we have a simple geometry or is the approximation to a simple geom-etry a limiting factor?

(c) Do we need to know the distance-dependence of the interaction or are we only really concerned with the effects when the surfaces are relatively close?

(d) Do we have the required material properties to hand?

The answers to each of these will set our strategy and so let us consider each in more detail.

It is only if we are carrying out very precise experiments to determine the attractive interaction by using model systems that we would need to tackle the full calculation. Often, we can use the value of the Hamaker constant as opposed to the distance-dependent Hamaker function. In these cases, the attractive energy equations for close distances would also suffice (for example, Equations (3.24) and (3.26)). The value of A could be calculated from Equa-tion (3.33) with the function $F(H/\lambda_e) = 1$. In addition, there are values for a wide variety of materials listed in the scientific literature [1, 10, 14, 15]. How-ever, it would be preferable to utilize Equation (3.33) rather than the combin-ing rules, as the values of $\varepsilon(0)$ and $n(0)$ are quite easy to find in the literature [16, 17]. Examples of this situation include problems of estimating adhesion of particles to surfaces, separation of particles during dispersion, estimation of the aggregation rates of particles, and estimation of the rheology of con-centrated aggregated systems.

If our materials are produced by techniques such as milling or high-speed dispersion, the particles may be small compact aggregates of primary crystallites or fragments of crystals. Plate-like surfaces may not be flat but 'stepped'. In cases like these, the simple approximations may be sufficient but the representation at close distances can easily be poor. The surface detail makes less difference at larger distances. We expect small liquid droplets to be spherical but we should be aware that the value of the interfacial tension is important in conjunction with the droplet size. In addition, at close distances of approach the surfaces are likely to flatten and increase the interaction area, thereby increasing the attraction.

At this point, it will be useful to look at an example calculation. For this purpose, we will take a poly(methyl methacrylate) latex particle in a simple aliphatic hydrocarbon, namely dodecane. This has been widely used as a model colloid by numerous workers. In this example, the particle size is 500 nm (i.e. a radius of 250 nm). The material properties that we need are:

$$\varepsilon_1(0) = 3.6; \; \varepsilon_3(0) = 2.01$$

$$n_1(0) = 1.492; \; n_3(0) = 1.411$$

$$\lambda_e = 99.9 \, \text{nm (from the speed of light/frequency}$$
$$\text{of the absorption peak in the UV)}$$

Equation (3.33) is used to give the Hamaker constant (i.e. putting $F(H/\lambda_e) = 1$) and the Hamaker function to show the effects of retardation. In addition, we use the full Hamaker expression for identical spheres (Equation (3.25b)) as well as the approximation for close approach (Equation (3.24b)). The results are shown in Figure 3.9, along with the plot for the Hamaker function.

The first point to note is that there is little retardation until a distance of $> 10\%$ of the wavelength of peak absorption. The drop has become significant at distances greater than 30 nm. Although the retardation is large at distances in excess of 500 nm, the attraction is becoming less than the thermal energy at $\sim 25\,^\circ\text{C}$ and we would be unable to discern the interaction, let alone the detail. The conclusion that we may draw from this is that the neglect of retardation and the use of the simplest equation for the interaction (Equation (3.24b)) will be adequate for most purposes for this system. If, for example, we had the task of dispersing these particles, the amount of work that we would put in would be the value of V_A at the closest distance of approach. Alternatively, if our task is to prevent the particles sticking firmly together, we have to supply another interaction which will have a greater potential energy but opposite in sign to the values at close distances.

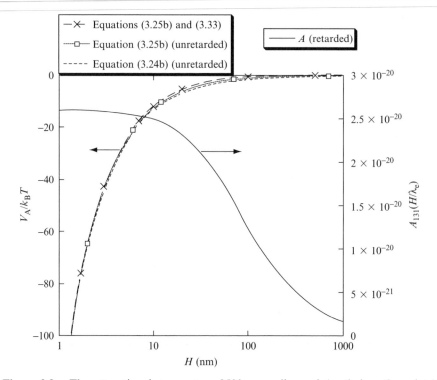

Figure 3.9. The attraction between two 2500 nm radius poly(methyl methacrylate) spheres in dodecane. The variation of the combined Hamaker constant with distance is also shown.

However, before we draw the conclusion that we can always take the simpler route, let us look at another system. This time, we will use water as the suspending medium with oil droplets as the particles. The oil phase here is tetradecane which represents a light fuel oil and has a chain length commonly found in nature. This chain length is the main fraction of coconut oil, for example. The particles are assumed to be $2\,\mu m$ and are all of the same size for our calculation. The material properties for this case are as follows:

$$\varepsilon_1(0) = 2.03;\ \varepsilon_3(0) = 80$$

$$n_1(0) = 1.418;\ n_3(0) = 1.333$$

$$\lambda_e = 103.4\,\text{nm}$$

In the example calculation, we are using Equations (3.33) and (3.25b) as we did in the previous example. Here though we are going to compare the

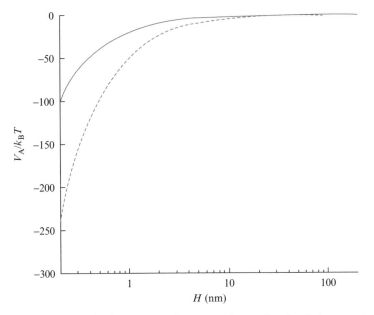

Figure 3.10. The attraction between two 2 μm tetradecane droplets in 'pure water' (----) and 'seawater' (—) (using Equations (3.25b) and (3.33)), where the low-frequency term is screened.

attraction between the oil drops in 'pure' water and 'seawater'. In the 'pure' water, the only ions are OH^- and H_3O^+, with concentrations of 10^{-7} mol dm^{-3} (i.e. no dissolved carbon dioxide). 'Seawater' is used to give such a high electrolyte concentration that we can assume that the low-frequency contributions to the attraction are completely screened out. We shall see later in this chapter how the ionic strength affects the range of electrical fields and that at high ionic strengths the range is only a nanometre or two.

The results are shown in Figure 3.10 and upon examination of this figure we should note that there is a large contribution due to the low-frequency term, that is, the first term on the right-hand side of Equation (3.33). This means that the oil droplets are 'less sticky' in seawater although an attraction of $\sim 100 k_B T$ is still a strong attraction. In most cases when we have particles dispersed in an aqueous medium, there is a significant ionic content. This may be from the addition of ionic surfactants, electrolytes (acids, bases or salts) and from ions associated with the particle surface due to surface charge. As we shall see later, the concentration is always at its greatest between pairs of particles and the neglect of the low-frequency contribution is an adequate approximation. The result now becomes similar to the case where the particles are dispersed in low-polarity media where the low-frequency contributions amount to $< 5\%$ of the Hamaker constant.

Finally, we should note that the case of the interaction between different particles is also quite tractable in an approximate form where we are assuming a dominant absorption peak in the UV. This is in a similar spirit to the earlier London model. However, the approximation we are using makes it easy to include a wide frequency variation so that retardation and the variation with dipolar materials are naturally included. The value of λ_e changes only a little from material to material and a value of 100 nm will normally suffice, although with aromatic materials such as benzene, toluene or polystyrene, 150 nm would give a better approximation. The equation for two particles of different materials, '1' and '2', acting across a medium, '3', is as follows [1]:

$$
A_{132} \approx \frac{3k_B T}{4} \left(\frac{\varepsilon_1(0) - \varepsilon_3(0)}{\varepsilon_1(0) + \varepsilon_3(0)} \right) \left(\frac{\varepsilon_2(0) - \varepsilon_3(0)}{\varepsilon_2(0) + \varepsilon_3(0)} \right)
$$
$$
+ \frac{3hc}{\lambda_e 3\sqrt{2}} \left\{ \frac{(n_1^2(0) - n_3^2(0))(n_2^2(0) - n_3^2(0))}{(n_1^2(0) + n_3^2(0))^{\frac{1}{2}}(n_2^2(0) + n_3^2(0))^{\frac{1}{2}} \left[(n_1^2(0) + n_3^2(0))^{\frac{1}{2}} + (n_2^2(0) + n_3^2(0))^{\frac{1}{2}} \right]} \right\}
$$

(3.36)

If the particles where spherical, Equation (3.25a) would be used to calculate the attraction. Although the equations look unwieldy, they are really quite straightforward to use in our calculations. Again, in aqueous systems the low-frequency contributions could be neglected and the following equation would provide an adequate approximation:

$$
A_{132} \approx \frac{3hc}{\lambda_e 3\sqrt{2}} \left\{ \frac{(n_1^2(0) - n_3^2(0))(n_2^2(0) - n_3^2(0))}{(n_1^2(0) + n_3^2(0))^{\frac{1}{2}}(n_2^2(0) + n_3^2(0))^{\frac{1}{2}} \left[(n_1^2(0) + n_3^2(0))^{\frac{1}{2}} + (n_2^2(0) + n_3^2(0))^{\frac{1}{2}} \right]} \right\}
$$

(3.37)

Metal particles such as gold, silver or platinum are important technologically as they are widely used in the form of inks for printing electronic circuit boards and integrated circuits. As they are conductors, the value of $\varepsilon_{Au}(0) = 8$. The characteristic frequency is the plasma frequency (i.e. $60\,\text{nm} < \lambda_e < 100\,\text{nm}$) and the value of the Hamaker constant for the metal particles in water is found to be about 4×10^{-19} J, i.e. metal particles are very much 'stickier' than particles made up of organic materials.

One final point that we must remember is that this and any other continuum approach is limited in terms of the minimum separation of surfaces. At very small distances, the molecular structures of both the surfaces and the species of the continuous phase become important. Even at a few tenths of a nanometre, we cannot expect the approximations to be reliable. At this dimension we are dealing with the size of hydrated ions, and solvent molecules such as aromatic or aliphatic hydrocarbons, as well as surface irregularities on what we might regard as 'smooth' surfaces. The problem of the dimensions of the intervening solution species leads us naturally to consider the next interaction that we will study.

9 THE DEPLETION INTERACTION

This is usually an attractive interaction that can cause aggregation of particles in concentrated dispersions although under some concentration conditions a weak repulsion may occur. It is most frequently considered to occur in dispersions to which non-adsorbing polymer has been added. This would be typically the situation where a polymer thickener had been added to a dispersion to control its rheological properties. A decorative paint system is a good example of this. Polymer thickeners are added to prevent sedimentation of pigment, as well as to give a thicker film and so increase the covering power. The particulate components of the paint (pigments and latex particles) are covered with adsorbed surfactants which restrict the adsorption of the soluble polymers. What we observe is that the thickening power of the polymer is much greater when particles are present in significant concentrations than when they are absent. However, the depletion interaction may also cause particles in the presence of concentrated surfactant micellar phases, or if there are two widely different size populations, to phase-separate.

The range of the interaction is no greater than the dimension of the solution species which is causing the effect, and so we can consider it to be a short-range interaction. That is, the range is much less than the dimensions of the particles that we are considering to represent the 'main population'. Although the greatest mass may be in the large particles that we dispersing, the greatest number is going to be in the small component which is causing the depletion interaction. For example, a few percent of a soluble, non-adsorbing polymer thickener could produce a marked interaction in a paint dispersion of 60–70 % solids. This interaction was first suggested by Asakura and Oosawa [18, 19] and further developed later [20, 21]. The concept is based on the idea of a 'depletion layer' close to the surface of a particle. Within this layer, the concentration of the non-adsorbing species (polymer molecules, for example) is lower than the average. This does not mean that there is complete absence of material in the layer but just that the centre of mass of the solution species cannot approach closer than the radius of gyration. This is shown in Figure 3.11, in which R_g is the radius of gyration.

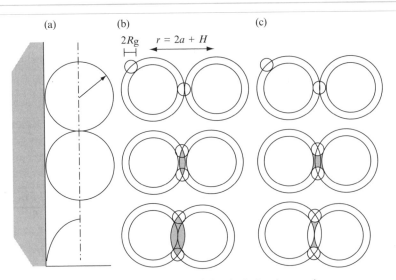

Figure 3.11. The origin of the depletion interaction.

As a starting point, think initially in terms of the non-adsorbing species consisting of hard spheres and then the concentration of material will follow the profile illustrated in Figure 3.11(a). Of course, if the material were a polymer in a good solvent, the increase in concentration would be slower, with the value calculated from the product of the profile shown and the concentration distribution within the coil. The mechanism for the interaction is a phase-separation of the large and small particles due to an effective decrease in concentration of the smaller particles as medium in the 'depletion layers' is squeezed out by the aggregation of the large particles. The argument then is as follows:

(1) There is a spherical shell around each particle which the centre of a small sphere cannot occupy, i.e. we should exclude this volume when we calculate the number density of small spheres, ρ_{p1}.
(2) The osmotic pressure from the small spheres is $\rho_{p1}k_BT$.
(3) As the large particles come together, the overlapping part of the depletion layers is now a volume which is freed from occupation by the small particles.
(4) The osmotic pressure acts to push the large particles together as there is an osmotic pressure difference between the volume excluded from the polymer and the bulk.
(5) The pressure multiplied by the interaction area is the force, and so we may construct the potential energy by integrating this force over the

separation distance. This gives the depletion contribution to the pair potential, V_d, simply as the osmotic pressure of the polymer solution multiplied by the overlap volume, v_o:

$$V_d = \left(\rho_{pl}k_B T\right)v_o \tag{3.38}$$

Figure 3.11(b) indicates the volume of liquid that is not accessible to a small sphere as the larger spheres come together (shaded portions in the figure). We may calculate this excluded volume from simple geometry. The lens is simply twice the volume of a spherical cap. The centre-to-centre separation of the large particles is r with the surface-to-surface separation H, the radius of the large particles is a, and that of the small particles is R_g. The volume of the cap of a sphere is therefore given by:

$$\text{volume of spherical cap of height } h = \frac{\pi}{3}h^2[3(a + R_g) - h] \tag{3.39a}$$

where:

$$h = \left[(a + R_g) - \frac{r}{2}\right] \tag{3.39b}$$

Now, we may write the depletion interaction between two spheres of equal size as:

$$V_d = -\left(\rho_{pl}k_B T\right)2\frac{\pi}{3}\left[(a + R_g) - \frac{r}{2}\right]^2\left[2(a + R_g) + \frac{r}{2}\right] \tag{3.40}$$

The negative sign indicates that the interaction is attractive. Multiplying this out, we have:

$$V_d = -(\rho_{pl}k_B T)\frac{4\pi}{3}(a + R_g)^3\left[1 - \frac{3}{4}\left(\frac{r}{a + R_g}\right) + \frac{1}{16}\left(\frac{r}{a + R_g}\right)^3\right] \tag{3.41a}$$

and if we write this in terms of the separation between surfaces, $H = r - 2a$, we have the following:

$$V_d = -(\rho_{pl}k_B T)\frac{4\pi}{3}(a + R_g)^3\left[1 - \frac{3}{4}\left(\frac{2a + H}{a + R_g}\right) + \frac{1}{16}\left(\frac{2a + H}{a + R_g}\right)^3\right] \tag{3.41b}$$

As we are assuming that the small particles are hard spheres, we may write the relative depletion potential in terms of the volume fraction of the small particles, φ_{pl}:

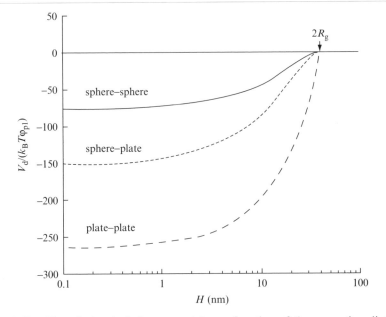

Figure 3.12. The relative depletion potential as a function of the separation distance between surfaces: R_g, 20 nm; a, 10^3 nm; plate overlap, 150 × 150 nm.

$$\frac{V_d}{\varphi_{p1} k_B T} = -\left(\frac{a + R_g}{R_g}\right)^3 \left[1 - \frac{3}{4}\left(\frac{2a + H}{a + R_g}\right) + \frac{1}{16}\left(\frac{2a + H}{a + R_g}\right)^3\right] \qquad (3.42)$$

This geometric function is plotted in Figure 3.12. The interaction is zero until a distance equal to the diameter of the small particle is reached between the large particles. A volume fraction of 0.05 would result in an attraction of $\sim 4k_B T$.

The interaction for other geometries is quite straightforward. For a sphere and a plate, we just have the osmotic pressure multiplied by the interaction area (the circular area of the cap of the sphere defined by the overlap of the depletion layers), integrated as we move the two surfaces closer together. The overlap distance is $(2R_g - H)$. This gives us the volume of a spherical cap and not a lens as was the case with two spheres. For plates, we have the osmotic pressure multiplied by the overlap area of the two plates to give the force. When this is integrated over distance as the plates are moved together, we obtain the volume of the two depletion layers multiplied by the osmotic pressure. Hence, for two plates the interaction increases linearly as the plates approach each other, with the overlap again being $(2R_g - H)$. So, when we write the expressions for models of the depletion potential for these geometries we have:

for a sphere with a plate,
$$\frac{V_d}{\varphi_{pl}k_B T} = \frac{1}{4}\left(2 - \frac{H}{R_g}\right)^2\left(1 + \frac{3a}{R_g} + \frac{H}{R_g}\right) \qquad (3.43)$$

and:

for two plates,
$$\frac{V_d}{\varphi_{pl}k_B T} = \frac{3A_o}{4\pi}\left(\frac{2}{R_g^2} - \frac{H}{R_g^3}\right) \qquad (3.44)$$

where the overlapping area of the two plates is A_o. These expressions are plotted in Figure 3.12, using some typical values for colloidal systems.

At this point, we have a simple model of the depletion potential, but we must remember that this is a *simplified* model and we should consider its limitations. The effect is of most interest with the case of solution polymers which have been added to dispersions, often as a rheology modifier or thickener. To exploit the effect, we need to take heed of the following:

(a) The polymer should be in a good solvent, i.e. $\chi < 0.5$.
(b) The polymer should be non-adsorbing, or if it adsorbs then the concentration must be sufficient to be greater than that required to saturate the surface, i.e. the concentration would ensure that we are on the plateau of an adsorption isotherm.
(c) We must remember that a polymer coil is not a hard sphere but is dynamic and with a concentration that decreases towards the periphery of the coil.
(d) The polymer concentration should be relatively high so that the osmotic pressure is large. This will invariably mean that we will have to use a more complicated expression for the osmotic pressure than the van't Hoff limit, i.e. $\Pi = \rho_{pl}k_B T + O\left(\rho_{pl}^2\right) + \ldots\ldots$

These last two points are important. The softness of a polymer coil can be expected to change the interaction profile. This possibility will be enhanced if the large particles are covered with a polymer layer. However, trying to model variations would not be very useful compared to the change that must occur due to the variation in coil size resulting from the polydispersity of the polymer. The molecular weight distribution of the majority of polymer thickeners is large and the variation in the value of R_g is, of course, directly related to this spread.

The osmotic pressure of the polymer solution can be tackled relatively easily. The simplest approach might be to determine it experimentally, although there are expressions in the literature which can be used. However, at this point we should look carefully at the model. When we do this, other factors become apparent. If we consider a single polymer coil in isolation, the dimension is a function of the molecular weight and the polymer/solvent interaction, i.e. the Flory–Huggins χ-parameter. We are not dealing with an isolated coil in pure solvent though and whatever else there is in solution can be expected to have an impact. Electrolytes increase the value of χ towards 0.5 for water-soluble polymers such as poly(ethylene oxide). Thus, we expect the polymer coil to shrink as the salt concentration is increased. Increasing temperature will also do this in aqueous systems as the H-bonding levels are reduced. The addition of low-molar-mass alcohols will produce a similar result. Surfactants can interact with polymer chains and change their spatial characteristics. The presence of other polymer molecules is particularly interesting. Of course, we have already drawn attention to the osmotic pressure change for the calculation of the depletion potential, but we should also remember that the dimensions of the polymer coils are also dependent on the polymer concentration. This is easy to visualize if we recall that at a concentration of c^*, the polymer coils would just fill the fluid volume. This means that at the peripheries the local polymer concentration becomes higher than that of an isolated coil. The diffusional motion increases this effect, thus resulting in a redistribution of the segments to give a more compact conformation. In other words, the value of R_g is reduced. Here, we have two competing effects: the increase in polymer concentration increases the depletion potential as the osmotic pressure is increased, although the dimensions of the depletion layer are reduced, thereby decreasing the strength of the interaction. The result is that we can find a maximum attraction at an intermediate polymer concentration.

A re-examination of Figure 3.11 should also indicate another problem with the model which can alter the profile of the potential curve. As two large particles approach at distances of $H < 2R_g$ we can see that there is a volume of liquid in between which does not contain the small spheres – the origin of the depletion effect. Note that we are not referring to the overlap volume of the depletion layers! That volume is just a mathematical consequence of integrating the interaction area with separation distance. The 'real' excluded liquid volume is a rather more complicated toroidal shape. However, this volume also varies as the interparticle distance varies. For an isolated particle pair in a large volume of polymer solution, there would be no change in the osmotic pressure. In a concentrated system, however, we must be careful. This volume variation exists with every interaction. At volume fractions > 0.3 with weakly aggregating systems like these, there can be six or more interactions per particle and this must be multiplied by the number of particles (divided by 2 so that we do not double the count, of course). What are the implications of this? We usually work in a region where the osmotic pressure is high and this also means that it is varying sharply with concentration. Reference to Figure 3.11(c) above shows clearly that the excluded volume goes through a maximum at intermediate separations. This means that in a concentrated system, the osmotic pressure is at a maximum at these intermediate separations also and is not a constant as the above model assumes. The result is that the maximum attractive force is not at contact but a little way prior to that point [22]. This is important for properties such as the rheology where, for example, the elasticity is a rate of change of the force with distance, that is, the curvature of the potential well.

REFERENCES

1. J. Isrealachvili, *Intermolecular and Surface Forces*, Academic Press, London, 1992.
2. L. D. Landau and E. M. Lifschitz, *Electrodynamics of Continuous Media*, Vol. 8, Pergamon, Oxford, UK, 1984.
3. A. D. McLachlan, *Proc. R. Soc. London, A*, **202**, 224 (1963).
4. A. R. Von Hippel, *Dielectric Materials and Applications*, Wiley, New York, 1958.
5. J. H. de Boer, *Trans. Faraday Soc.*, **32**, 10 (1936).
6. H. C. Hamaker, *Physica*, **4**, 1058 (1937).
7. B. V. Derjaguin, *Kolloid Z.*, **69**, 155 (1934).
8. D. Langbein, *Theory of van der Waals Attraction*, Tracts in Modern Physics, Springer, Berlin, 1974.
9. R. J. Hunter, *Foundations of Colloid Science*, Vol. I, Oxford University Press, Oxford, UK, 1987.
10. W. B. Russel, D. A. Saville and W. R. Schowalter, *Colloidal Dispersions*, Cambridge University Press, Cambridge, UK, 1989.
11. J. Mahanty and B. W. Ninham, *Dispersion Forces*, Academic Press, London, 1976.
12. P. W. Atkins, *Physical Chemistry*, 2nd Edn, Oxford University Press, Oxford, UK, 1982.
13. V. A. Parsegian and G. H. Weiss, *J. Colloid Interface Sci.*, **81**, 285 (1981).
14. J. Gregory, *Adv. Colloid Interface Sci.*, **2**, 396 (1973).
15. D. B. Hough and L. R. White, *Adv. Colloid Interface Sci.*, **14**, 3 (1980).
16. R. C. Weast and M. J. Astle (Eds), *Handbook of Chemistry and Physics*, 60th Edn, CRC Press, Boca Raton, FL, USA, 1981.
17. J. Brandrup and E. H. Immergut (Eds), *Polymer Handbook*, 2nd Edn, Wiley, New York, 1975.
18. S. Asakura and F. Oosawa, *J. Chem. Phys.*, **22**, 1255 (1954).
19. S. Asakura and F. Oosawa, *J. Polym. Sci.*, **33**, 183 (1958).
20. A. Vrij, *Pure Appl. Chem.*, **48**, 471 (1956).
21. J. F. Joanny, L. Liebler and P.-G. de Gennes, *J. Polym. Sci. Polym. Phys. Ed.*, **17**, 1073 (1979).
22. J. W. Goodwin, R. W. Hughes, H. M. Kwaamba and P. A. Reynolds, *Colloids Surf. A*, **161**, 361 (2000).

Chapter 4

Forces of Repulsion

1 INTRODUCTION

In Chapter 3, we discussed the attractive interactions between colloidal particles. There are other contributions to the pair potential which prevent particles from adhering to each other. These are the forces of repulsion and this chapter will describe the most important of these.

2 ELECTROSTATIC INTERACTIONS

Many interfaces in aqueous systems carry an electrical charge. Like charges repel each other. In the simplest case, this may be described by Coulomb's law; however, we shall see that the ionic content of the system modifies this. The repulsion though can be strong and prevent interfaces from coming together. Let us consider a soap film, as an example. In this case, a surfactant is strongly adsorbed at the interface with hydrocarbon tails on the vapour side and ionic head groups on the water side of each interface. Sodium or potassium fatty acid salts make the classic soaps with the number of carbon atoms ranging from 14 to 18, depending on the source. When a foam is produced, the water drains due to gravity and the close-packed arrays of ionic end groups approach each other. At some point, the forces come into balance with the magnitude of the electrostatic repulsion controlling the final thickness of the foam films. This repulsion occurs between any 'like-charged' interfaces and is an important factor in determining the colloidal behaviour of aqueous systems and also those of intermediate polarity (those with a relative permittivity of 10).

Colloids and Interfaces with Surfactants and Polymers – An Introduction   J. W. Goodwin
© 2004 John Wiley & Sons, Ltd   ISBN: 0-470-84142-7 (HB)   ISBN: 0-470-84143-5 (PB)

The model that we will use is mainly a continuum model in which we will assume any charged surface to be uniformly charged and therefore structureless. Our colloidal systems will also be electrically neutral so that any surface charge is always balanced by an equal amount of charge opposite in sign in the continuous phase. The distribution of the solution charge close to the charged surface though is where we put our efforts. The two layers of charge are known as the *electrical double layer*. Ions in solution of the same sign as the charged surface are referred to as *co-ions* and those of opposite sign as *counter-ions*. There will always be some ions present in our aqueous solvent phase, of course, such as H_3O^+ and OH^-, as well as species such as carbonate, silicate, etc., even when we have not intentionally added electrolytes. Ions which interact specifically with the surface to vary the value of the surface charge are known as *potential-determining ions*; H_3O^+ would be an example of such an ion with our fatty acid soap films, for example, as the carboxylic acid is a weak acid and the degree of dissociation would be different at pH 2 and pH 10. Electrolytes whose ions play no direct part in the charging mechanism are known as *indifferent electrolytes*. However, the specificity of the ion–surface interaction is of major importance in controlling the behaviour of colloidal systems and is still a rich field of study.

3   THE ORIGINS OF SURFACE CHARGE

There are several mechanisms by which an interface may acquire a charge. In many cases, we have some control over the value of the charge. Each mechanism is introduced below with some examples but it is important to be aware that there can be more than one mechanism operating in our practical systems.

3.1   Adsorbed Layers

Surfactants are often added as a component to disperse systems, for example, as a wetting aid for the dispersion of dry powders, and as an emulsifier for liquid systems, as well as a stabilizer for the end product. Ionic surfactants are relatively cheap and are frequently added. The final conformation at the interface is for the hydrophobic tails to be at the interface with the ionic groups outermost. The surface density is always high in a surfactant-stabilized system. Proteins also adsorb onto hydrophobic surfaces and provide an hydrophilic charged outer layer. An example is provided by the fat droplets in milk which are stabilized by casein. The charge is negative as it results from the carboxyl groups on the protein.

  Natural rubber latex is a dispersion of poly(*cis*-isoprene) with a negatively charged surface from the adsorbed protein on the latter. Reduction of the pH prevents the dissociation of the carboxyl groups and coagulation results. Synthetic rubber latices (poly(styrene–butadiene) rubbers) have some charged groups chemically bound to the polymer but invariably have a high concentration of anionic surfactants adsorbed on the surfaces of the particles. The surfactant has two functions: first, as an emulsifier for the monomer prior to synthesis, and secondly to provide colloid stability to the final product.

Poly(acrylic acid) can be used with oxides in order to provide a coated surface with a negative charge at pH values where the oxide surface may have a low positive charge. Inorganic 'macro-ions' can also be used for this purpose. China clay (kaolinite) is an important additive in paper to provide a glossy surface and it can be stabilized with poly(acrylic acid) adsorbed onto the positively charged edges. 'Calgon', a phosphate which adsorbs on the faces, is also used as a stabiliser for kaolinite. Titanium dioxide particles are often coated with a layer of alumina and poly(acrylic acid) will adsorb onto the surface at neutral pH.

3.2 Ionogenic Surfaces

The charge on this type of surface is controlled by the ionization of chemical moeities at the surface. Carboxyl groups that are chemically bound to the polymers of synthetic latices provide an example. The charge is a function of pH as the degree of dissociation itself is a function of pH. Although the pK_a of an isolated –COOH group is equivalent to a pH of ~ 4, this is not the situation with a surface with many groups in close proximity. The dissociation of one group makes it more difficult for the immediate neighbours to dissociate. This is the *polyelectrolyte effect* and means that the surface has a variable pK_a and pH values as high as 9 may be required to ensure dissociation of all of the surface groups.

The surfaces of uncoated oxide particles have surface hydroxyl groups. At high pH, these can ionize to give $-O^-$ and at low pH the lone pair on the oxygen can hold a proton to give $-OH_2^+$. For example, if we consider the surface of a titanium dioxide particle we have oxygen atoms on the surface of the crystal giving an amphoteric surface with the scheme:

$$\text{Surface } TiOH_2^+ + H_2O \xleftarrow{\;H_3O^+\;} \text{Surface } TiOH \xrightarrow{\;OH^-\;} \text{Surface } TiO^- + H_2O$$

Hence, this type of surface not only shows a variation in the magnitude of the surface charge with pH but also a variation in the sign. At an intermediate pH value, the charge can be reduced to zero and the pH at which this occurs is termed the *isoelectric point* (iep) or the *point of zero charge* (pzc). Frequently, the two terms are used interchangeably but they are not always the same. For a simple oxide surface, the former term is to be preferred as we

Table 4.1. Values of the isoelectric points for a number of oxides

Oxide	Formula	iep
α-Alumina	α-Al_2O_3	9.1
Haematite	α-Fe_2O_3	6.7
Magnesium oxide	MgO	12.5
Rutile	TiO_2	3.5
Silica	SiO_2	2.0
Zinc oxide	ZnO	10.3
Zircon	ZrO_2	5.7

expect the charge to be uniformly zero across the surface. With rather more complex materials, where there is more than one type of surface present, we may have a situation where both surfaces carry a charge, but of opposite sign. Then, at some pH value we can have zero net charge as the two values balance each other. This would then be the pzc. China clay (kaolinite) is a layer lattice aluminosilicate, where the edges are positive with negative basal surfaces. In this case, the iep of the edges is at pH $\sim$ 8 with the face iep at pH $\sim$ 2 and a pzc for the particle at pH $\sim$ 6. Table 4.1 lists the isoelectric points for some oxides.

3.3 Isomorphous Substitution

This is a common occurrence in clay minerals. The basic structure of a clay particle is an aluminosilicate layer lattice. As the clay is formed, it crystallizes with a layer of silicon atoms tetrahedrally coordinated to oxygen atoms (an SiO_2 layer). The next layer of the lattice is aluminium with octahedrally coordinated oxygens (an Al_2O_3 layer), some of which are shared with the tetrahedral silica layer. This layer structure is repeated throughout the crystal. If it is a 1:1 layer lattice structure, there are alternating layers of silica and alumina (e.g. kaolinite). Alternatively, it can crystallize in a 2:1 structure, with the alumina layer sandwiched between two silica layers (e.g. montmorillinite). During the crystallization process, an occasional silicon atom can be substiututed by an aluminium atom, and more frequently an aluminium atom by a magnesium atom. A *small* amount of such substitution does not produce too much distortion of the lattice to stop it growing and it retains the same structure – hence, the term *isomorphous substitution*. Thus, the number of oxygen atoms is the same. As anions they are larger than the cations in the structure and their packing is the dominant factor. However, the difference in valency of the magnesium, compared to the aluminium anions, for example, means that part of the oxygen anion coordination will be incomplete, thus resulting in a negative charge. This charge is balanced by soluble cation species at the surface of the crystal, with the result that the basal surfaces of clay minerals carry a significant negative charge. In the dry state, the counter-ions to this surface charge are located on the surface, while in the hydrated state they are in solution near the surface. The 2:1 layer lattice clays have a unit cell ~ 1 nm thick and on hydration, water penetrates between the layers and the negative charges of the surfaces repel and cause the clays to swell. This swelling can result in a separation between (orginal) unit cells of a greater dimension than that of the (original) unit cell. This means that the expansion of the clay can be very marked. Such a phenomenon causes major difficulties in building on land with high clay contents.

3.4 Differential Solution of Surface Ions

When the colloidal particles are made up of sparingly soluble salts, dissolution occurs until the concentrations of the ionic components in solution correspond to the solubility products of the compounds. Silver halides are a much studied example of this class of material, with silver bromide dispersions, for example, having a long history of a commercially important component of photographic emulsions. For many years, silver iodide was used as a model colloid for academic studies. It is readily produced from mixing silver nitrate and potassium iodide solutions. Under the correct conditions, a dispersion is formed with the particles confined to a narrow size distribution and with a cubic shape. Now, the solubility product, $K_s = [Ag^+][I^-] = 10^{-16}$. If the precipitation is produced under conditions which have an excess of I^-, say 1×10^{-4} M, we form a dispersion with negatively charged particles. Using $p[I^-]$ as $-\log[I^-]$, we have a solution with a $p[I^-]$ of 4, and hence the $p[Ag^+] = 12$, and the solution of Ag^+ ions is suppressed with the surface of the particle consisting of I^- species. At $p[I^-] = 10.5$, i.e $p[Ag^+] = 5.5$, the surface populations of the two ion types are equal and the pzc is reached as the net charge is zero. Here, the potential-determining ions are Ag^+ and I^-. We might have expected that the pzc would have occurred when there were equal numbers of Ag^+ and I^- species present, i.e. at $p[I^-] = 8$; however, the solvation of the smaller cation is greater and the adsorption is in favour of the large anion.

On the surface of the silver iodide crystal, at or close to the pzc, we have a large number of charged species. This means that as the surface is charged up only a relatively small increase in the number of anions or cations is required and we may represent the surface potential as we would for an electrode surface by the Nernst equation [1]:

$$\psi_0 = \frac{RT}{F_c} \ln \left(\left[\frac{Ag^+}{[Ag^+]_{pzc}} \right] \right) \qquad (4.1)$$

where F_c is the Faraday constant. This gives a surface potential of $\sim \pm 60\,mV$ for plus or minus a factor of 10 in the silver ion concentration from that corresponding to the pzc.

Other colloidal crystallites can also show *Nernstian behaviour* and some examples are given by Hunter [1]. However, it should be pointed out that this behaviour is far from universal for colloidal particles. For example, with amphoteric surfaces such as oxides, the surface charge population close to the pzc is very small and as the surface charges up the activity of the species at the surface changes and we no longer find Nernstian behaviour. There are a variety of models available for different types of surfaces and a good summary can again be found in the text by Hunter [1].

3.5   The Structure of the Electrical Double Layer

When we have a colloidal particle with a charged surface there is always an equal and opposite charge in the solution. The structure of the solution part of the double layer must now be considered. We will start with modelling a flat surface and then consider curved interfaces. The earliest treatment was due to Helmholtz and this described a simple model of a uniformly charged surface with ions of opposite charge, the counter-ions, in a uniform layer adjacent to the surface and treated as point charges. However, a more complex picture superseded this model. The first point that we need to consider is that the solvated counter-ions have a finite size, that they can interact laterally and that there may be specific chemical interactions with the surface. That is, we must be aware that other than straightforward electrostatic forces may be involved. The second point to remember is that the diffusive motion of the ions will oppose the tendency to concentrate the counter-ions in the interfacial region and result in a diffuse array. It is in studying this diffuse region that we will put most of our effort here after first considering the first layer of ions.

❑   THE STERN PLANE

This is the inner region of the solution part of the double layer. It is usually modelled by using the *Langmuir isotherm*, which describes the formation of a monolayer. We should be familiar with the isotherm in the context of gas adsorption where the adsorption is a function of the pressure and the fraction of the sites that are occupied. In our present case, the adsorption energy is made up of the electrostatic attraction and any other 'specific' interactions. So, we have a monolayer consisting mainly of counter-ions at the surface whose population is a function of the electrostatic potential plus specific chemical interactions, as well as the ionic content of the aqueous phase. (Here, the ionic strength is analogous to pressure in the gas-phase adsorption.) Another problem is that in colloidal systems the molecular structure of the surface means that the discrete nature of the charges needs to be considered at this level. This acts to increase the site occupation so that, if we were to estimate the adsorption energy, a lower energy would be required to attain the same level of counter-ion density at the surface than would be the case if the surface charge is considered to be 'smeared out' uniformly over the surface.

The picture that is emerging now is a complicated one and a detailed analysis is outside the scope of this introductory text. Good descriptions of the current ideas are given in the texts by Hunter [1] and Lyklema [2], along with the major references to the original research papers. For our present purposes, we will consider our surface to be uniformly charged with a surface charge density of σ_0 C m^{-2} and adjacent to this a layer of counterions. It is to the outer edge of this layer that we must now turn our attention. At this plane, the adsorbed ions have changed the charge to σ_δ and the potential relative to ground (i.e. at a very large, and so effectively infinite, distance from the surface) is ψ_δ. This is just the work done in bringing a point charge from infinity to this plane. From here, we will be treating the ions as point-charges. Now, unfortunately this is a difficult quantity to measure on a routine basis but we can relatively easily estimate a potential from electrokinetic measurements where we have a motion between the fluid and the interface. This potential is known as the ζ-potential and is the potential where the centre of the first layer of solvated ions that are moving relative to the surface is located. This is termed the 'shear plane', but it is not at a well defined distance from the surface and so we have the situation where we can determine a potential but where the position is a little uncertain at ~ 0.5 nm or so from the surface. This distance would be about three times the radius of a solvated ion, but the utility of the ζ-potential is that it reflects the value of ψ_δ which can differ in sign as well as markedly in magnitude from the potential at the surface, ψ_0. An illustration of our emerging model of the double layer is given in Figure 4.1.

The population of the Stern layer is a function of the ion type. Multivalent ions bind more strongly than monovalent ones. The solvation and polarizability are also important factors. These characteristics lead to the specificity of the various ions. So, for example, with a positively charged surface we observe a very marked difference in the binding of the halide anions in the series F$^-$ through to I$^-$. The addition of multivalent counter-ions to a colloidal dispersion can result in a densely populated Stern layer and a reversal in the sign of the charge relative to the particle surface.

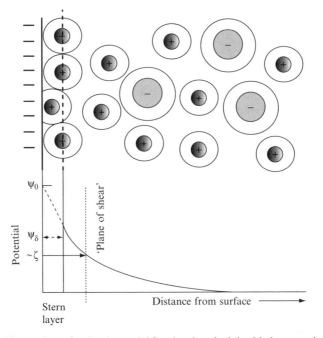

Figure 4.1. Illustration of a simple model for the electrical double layer at the surface of a colloidal particle.

☐ THE DIFFUSE DOUBLE LAYER

This is the part of the electrical double layer outside the Stern layer where the concentration of ions is determined by the competing effects of electrostatics and diffusion. What we are going to do is to calculate the distribution of ions, and from this the potential as a function of distance. Ultimately, this will enable us to model the electrostatic repulsive energy as we estimate the work done in bringing one surface up to a second surface.

The model that we are going to use is known as the *Gouy–Chapman model*, named after the two physicists who developed it independently. We start by defining the potential at a distance x from our surface as $\psi(x)$. As we are considering a flat surface which is uniformly charged, there will only be a dependence in the x-direction, with the potential being constant in planes parallel to our surface. In addition, the ions are treated as though they were point-charges and the number of ions of 'type i' per unit volume in the bulk electrolyte is n_{i0}. We make use of the Boltzmann distribution to estimate the ion density at x:

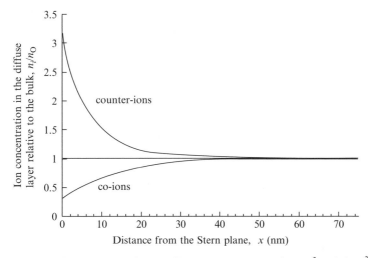

Figure 4.2. Local ion concentration profiles: $\psi(0) = -30\,\text{mV}$, in $10^{-3}\,\text{mol dm}^{-3}$ NaCl.

$$n_i = n_{i0} \exp\left(-\frac{z_i e\,\psi(x)}{k_B T}\right) \tag{4.2}$$

where e is the charge on the electron and z_i is the valency of the ion. We need to be careful about signs here and remember that as the sign of the charge on the counter-ion is always opposite to that of the surface, the exponent will always be negative for the counter-ion concentration and positive for the co-ion concentration. Equation (4.2) describes the situation where the counter-ion concentration increases close to the interface – adsorbed – while the concentration of the co-ions is reduced close to the interface – negatively adsorbed. Figure 4.2 shows how the local ion concentration profiles behave according to Equation (4.2). The number of charges per unit volume, i.e. the space charge density, ρ, is given by:

$$\rho = \sum_i n_i z_i e \tag{4.3}$$

In order to make the algebra more transparent, we can take the case of a symmetrical electrolyte where $z_- = -z_+$. Using Equation (4.2), we can write the net space charge density as:

$$\rho = n_0 z e \left[\exp\left(-\frac{z e\psi(x)}{k_B T}\right) - \exp\left(\frac{z e\psi(x)}{k_B T}\right)\right] = -2n_0 z e \sinh\left(\frac{z e\psi(x)}{k_B T}\right) \tag{4.4}$$

recalling that the identity $\sinh(x) = (e^x - e^{-x})/2$.

Now, Poisson's equation relates the space charge density to the curvature of the potential, as follows:

$$\nabla^2 \psi(x) = -\frac{\rho}{\varepsilon_0 \varepsilon_r} \tag{4.5}$$

where the Laplace operator $\nabla^2 = \partial^2/\partial x^2 + \partial^2/\partial y^2 + \partial^2/\partial z^2$, while for a planar surface, $\nabla^2 = d^2/dx^2$.

Substituting for ρ in Equation (4.5) gives us the Poisson–Boltzmann equation, as follows:

$$\nabla^2 \psi(x) = \frac{2n_0 z e}{\varepsilon_0 \varepsilon_r} \sinh\left(\frac{z e \psi(x)}{k_B T}\right) \tag{4.6}$$

Rewriting this for the planar interface:

$$\frac{d^2 \psi}{dx^2} = \frac{2n_0 z e}{\epsilon_0 \epsilon_r} \sinh\left(\frac{z e \psi(x)}{k_B T}\right) \tag{4.7}$$

Prior to seeking solutions to this equation, let us specify the boundary conditions that we have:

$$x = \delta \qquad\qquad x = \infty$$

$$\psi(x) = \psi_\delta \qquad \psi(x) = 0$$

$$\frac{d\psi(x)}{dx} = 0 \tag{4.8}$$

As an initial proposition, let us look at the case for small potentials where $(z e \psi(x)/(k_B T)) < 1$. For a univalent electrolyte, we have $z = 1$ and then we have the condition that $\psi(x) < 25\,\text{mV}$. When x is small, $\sinh(x) \sim x$ and Equation (4.7) becomes:

$$\frac{d^2 \psi}{dx^2} \approx \frac{2n_0(z e)^2}{\varepsilon_0 \varepsilon_r k_B T} \psi(x) \tag{4.9}$$

This is known as the Debye–Hückel approximation. Equation (4.9) is a linear homogeneous second-order differential equation whose solution is in terms of exponentials and with the boundary conditions given above we have the potential as a function of distance away from the Stern layer as:

$$\psi(x) \approx \psi_\delta \exp(-\kappa x) \tag{4.10a}$$

where for any electrolyte:

$$\kappa = \left[\frac{\sum_i \left(e z_i^2 \right) n_0}{\varepsilon \varepsilon_0 k_B T} \right]^{\frac{1}{2}} \tag{4.10b}$$

In Equation (4.10b), κ is the Debye–Hückel parameter that we find in the theory of electrolytes and which controls the rate of decay of potential with distance away from a surface.

We will look in more detail shortly at the behaviour of κ as the electrolyte concentration and type are changed, but first we should consider the solution of Equation (4.7) at higher potentials. With a little manipulation (see Hunter [1] and Russell *et al.* [3], for example), Equation (4.7) may be integrated to give:

$$\psi(x) = \frac{2k_B T}{ze} \ln \left[\frac{1 + \exp\left(-\kappa x\right) \tanh\left(\dfrac{ze\psi_\delta}{4k_B T} \right)}{1 - \exp\left(-\kappa x\right) \tanh\left(\dfrac{ze\psi_\delta}{4k_B T} \right)} \right] \tag{4.11}$$

recalling that the identity $\tanh(x) = (e^x - e^{-x})/(e^x + e^{-x})$, and so $= (e^{2x} - 1)/(e^{2x} + 1)$. A comparison of the results for surfaces with potentials of $-25\,\text{mV}$ and $-75\,\text{mV}$ are shown in Figure 4.3, calculated from Equations (4.10) and (4.11). The 1:1 electrolyte concentration in this case was 10^{-3} M.

Let us now consider the case where we have a high potential ($\tanh(x) \sim 1$ as $x \gg 1$). At long distances from the surface, x is large and $\exp(-\kappa x)$ is

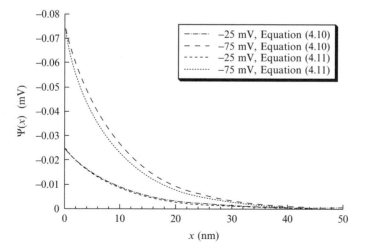

Figure 4.3. Decrease of electrical potential with distance for a planar surface in $10^{-3}\,\text{mol dm}^{-3}$ NaCl.

small and so the logarithms can be expanded as a series and only the leading terms used. Hence at long distances, we have:

$$\psi(x) \sim \frac{4k_B T}{ze} \tanh \left(\frac{ze\psi_\delta}{4k_B T} \right) \exp\left(-\kappa x \right) \tag{4.12}$$

So, we have the exponential fall of potential at large distances and, as the tanh function $\rightarrow 1$ for high potentials, the behaviour appears to be that of a surface with a reduced potential $ze\psi/(k_B T) \sim 4$. Therefore, at long distances our surface appears to have a potential of 100 mV regardless of how much higher it may be. This is important as we are often interested in the interactions of highly charged surfaces as they approach at separations corresponding to $\kappa x \gg 1$.

The Debye–Hückel parameter characterizes the decay of the potential with distance from the surface. Equation (4.10b) shows that this is a function of both the concentration of the bulk electrolyte and the valency of the ions. Figure 4.4 gives plots for various electrolyte types. Thus, when $\kappa x = 1$ the potential will have fallen to 37 % of the value at the start of the diffuse layer. For the potential to have fallen to ~ 1 % of the Stern layer value, the distance will have to be as far away from the surface as ~ 4.5 'decay lengths'. Hence, even at moderate electrolyte concentrations we can see that the decay of the potential occurs at distances comparable with the dimensions of many colloidal particles.

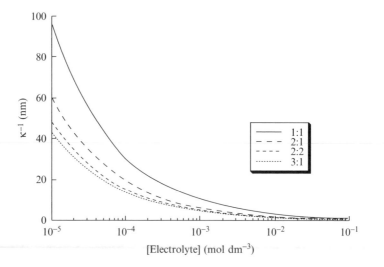

Figure 4.4. The Debye–Hückel decay length as a function of the concentration and type of electrolyte.

When the ζ-potential was introduced it was defined as the potential at the 'shear plane' and was taken as an approximation to the potential of the Stern layer, i.e $\psi_\delta \sim \zeta$, although the location of the shear plane is about a hydrated ion diameter further from the surface. Because it is often easy to measure, it is normally the ζ-potential that we will use for calculating diffuse layer interactions. It is straightforward to estimate the charge at the Stern layer which is, of course, equal but opposite in sign to the diffuse layer charge. We can use Equation (4.5) to estimate the charge at a distance δ from the surface:

$$\sigma_\delta = -\int_\delta^\infty \rho \, dx = \varepsilon_0 \varepsilon_r \int_\delta^\infty \frac{d^2\psi(x)}{dx^2} dx \qquad (4.13)$$

so that:

$$\sigma_\delta = \varepsilon_0 \varepsilon_r \left(\frac{d\psi(x)}{dx}\right)_\delta^\infty \qquad (4.14)$$

Our boundary condition is that at infinity the slope of the potential–distance curve is zero, and so we then have:

$$\sigma_\delta = -\varepsilon_0 \varepsilon_r \frac{d\psi_\delta}{dx} \qquad (4.15)$$

Equation (4.9) gives us the distance-dependence of the potential for small potentials, with the slope at δ given by:

$$\left.\frac{d\psi(x)}{dx}\right|_\delta = -\kappa\psi_\delta \exp(-\kappa\delta) = -\kappa\psi_\delta, \text{ as } \kappa\delta \to 0 \qquad (4.16)$$

and so:

$$\sigma_\delta = \kappa\varepsilon_0 \varepsilon_r \psi_\delta \approx \kappa\varepsilon_0 \varepsilon\zeta \qquad (4.17)$$

For higher potentials, we can use the same route but now using Equation (4.7) instead of Equation (4.10) to yield [1]:

$$\left.\frac{d\psi(x)}{dx}\right|_\delta = -\frac{2\kappa k_B T}{ze} \sinh\left(\frac{ze\psi_\delta}{k_B T}\right) \qquad (4.18)$$

so that:

$$\sigma_\delta \approx \frac{2\kappa\varepsilon_0\varepsilon_r k_B T}{ze} \sinh\left(\frac{ze\zeta}{2k_B T}\right) \qquad (4.19)$$

❏ THE DIFFUSE LAYER AND SPHERICAL PARTICLES

The solution of the Poisson–Boltzmann equation is not available in an analytical form except when the Debye–Hückel approximation for low potentials is used. As there is spherical symmetry it is usual to use spherical polar coordinates instead of Cartesian so that the equivalent form of Equation (4.7) for a symmetrical electrolyte with r as the distance from the centre of the particle is:

$$\frac{1}{r^2}\frac{\partial}{\partial r}r^2\frac{\partial\psi(r)}{\partial r} = \frac{2zen_0}{\varepsilon_0\varepsilon_r}\sinh\left(\frac{ze\psi(r)}{k_B T}\right) \qquad (4.20)$$

With the Debye–Hückel approximation, we obtain the diffuse layer potential as a function of distance (as for a particle of radius a):

$$\psi(r) = \psi_\delta\frac{a}{r}\exp\left[-\kappa(r-a)\right] \qquad (4.21)$$

Here, we are assuming δ to be very much smaller than a. We can also estimate the Stern layer charge along similar lines to the flat surface, to give:

$$\sigma_\delta = \frac{\varepsilon_0\varepsilon_r\psi_\delta(1+\kappa a)}{a} \qquad (4.22)$$

In addition, the charge at the 'shear plane' can be estimated if the ζ-potential is substituted for the Stern potential.

For potentials in excess of $25\,\text{mV}$, we need numerical solutions to the Poisson–Boltzmann equation. Loeb *et al.* [4] have provided extensive tables from such numerical calculations and these authors, as well as others, have also supplied some approximate analytical expressions (see, for example, Hunter [1] and Oshima *et al.* [5]). As an example, a useful expression for the charge of the Stern layer, approximating it to that at the 'shear plane', would be [4]:

$$\sigma_\delta \approx \frac{\varepsilon_0\varepsilon_r\kappa k_B T}{ze}\left[2\sinh\left(\frac{ze\zeta}{2k_B T}\right) + \frac{4}{\kappa a}\tanh\left(\frac{ze\zeta}{4k_B T}\right)\right] \qquad (4.23)$$

4 THE INTERACTION BETWEEN DIFFUSE DOUBLE LAYERS

As two charged surfaces approach each other, they interact. If the sign of the charge is the same, they usually repel each other; alternatively, if the charges are opposite, they attract. In a vacuum, the repulsion is described by Coulomb's law but in a liquid medium the interaction is screened by the ion atmosphere. The algebra for two planar surfaces is easiest to follow. The model used to calculate the repulsion uses the ion concentration between the surfaces to give an osmotic pressure difference between the surfaces and the bulk electrolyte. The potential distribution is assumed to be the sum of the potentials due to the two surfaces. This is illustrated in Figure 4.5, where we plot the potential distribution between two flat surfaces with a small overlap of diffuse layers.

4.1 The Interaction between Similar Flat Plates

The simplest route is to estimate the osmotic pressure at the mid-plane position between two surfaces separated by a distance h i.e. at $h/2$ where $d\psi(x)/dx = 0$ and $\psi(x) = \psi_m$. Once we have the osmotic pressure, the potential energy per unit area of surface is calculated from an integration of the force (osmotic pressure) we overcome in bringing the two surfaces from an infinite distance apart to position H.

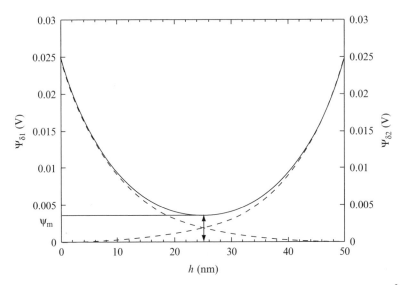

Figure 4.5. The potential profile between two planar surfaces at 25 mV, in 10^{-3} mol dm^{-3} NaCl.

The difference in the ion concentration at the mid-plane and the bulk electrolyte gives us the osmotic pressure, $\Pi(H)$:

$$\Pi(H) = k_B T(n_+ + n_- - 2n_0) \qquad (4.24)$$

Using the Boltzmann distribution of ions (Equation (4.2)), we can rewrite Equation (4.24) for a symmetrical electrolyte as:

$$\Pi(H) = 2n_0 k_B T \left[\cosh \left(\frac{ze\psi_m}{k_B T} \right) - 1 \right] \qquad (4.25)$$

recalling that the identity $\cosh(x) = (e^x + e^{-x})/2$. For small values of x, we can expand $\cosh(x) \sim 1 + x^2/2$, and so for small potentials, we have:

$$\Pi(H) \approx \frac{\kappa^2 \varepsilon_0 \varepsilon_r}{2} \psi_m^2 \qquad (4.26)$$

Substituting for the mid-plane potential, $\psi_m = 2\psi(H/2)$, we now have:

$$\Pi(H) \approx \frac{\kappa^2 \varepsilon_0 \varepsilon_r}{2} [2\psi_\delta \exp(-\kappa H/2)]^2 = 2\kappa^2 \varepsilon_0 \varepsilon_r \psi_\delta^2 \exp(-\kappa H) \qquad (4.27)$$

and the potential energy is:

$$V_R = - \int_\infty^H \Pi(H) dH \qquad (4.28)$$

which, with the boundary condition that $\Pi(H) \to 0$ as $H \to \infty$, gives:

$$V_R = 2\kappa \varepsilon_0 \varepsilon_r \psi_\delta^2 \exp(-\kappa H) \qquad (4.29a)$$

namely:

$$V_R \approx 2\kappa \varepsilon_0 \varepsilon_r \zeta^2 \exp(-\kappa H) \qquad (4.29b)$$

The same route is taken for systems with higher potentials where we cannot justify the simplifying Debye–Hückel approximation. If we look at the case of a weak overlap of the diffuse layers so that $\kappa H > 1$, we can still simply add the local potentials that we estimate from the isolated surfaces (Equation (4.12)). Following the same route as used above for the low-potential case, we can derive the resulting expression for V_R:

$$V_R = \frac{64 n_0 k_B T}{\kappa} \tanh^2 \left(\frac{z e \psi_\delta}{4 k_B T} \right) \exp\left(- \kappa H \right) \tag{4.30}$$

4.2 The Interaction between Dissimilar Flat Plates

Of course, the interaction will not always be between two identical surfaces. For two flat plates with potentials $\psi_{\delta 1}$ and $\psi_{\delta 2}$, we have, in the Debye–Hückel limit [1]:

$$V_R = \frac{\varepsilon_r \varepsilon_0 \kappa}{2} \left\{ \left(\psi_{\delta 1}^2 + \psi_{\delta 2}^2 \right) [1 - \coth{(\kappa H)}] + 2 \psi_{\delta 1} \psi_{\delta 2} \operatorname{cosech}{(\kappa H)} \right\} \tag{4.31}$$

Some results using Equation (4.31) are plotted in Figure 4.6 for two surfaces with potentials of -25 and -15 mV. In addition, the results obtained for two surfaces of equal potential at 20 mV are shown for reference. We see a maximum in the interaction energy at $\kappa H \sim 0.5$. At close separations, the potential energy decreases and eventually becomes attractive. At the maximum of the curve, the electrostatic interaction force is zero and is attractive in the above example when $\kappa H < 0.5$. This is quite an exciting result as it indicates why some materials with a similar sign charge may stick together or *heterocoagulate*. For example, we could picture a protein-coated surface of a cell

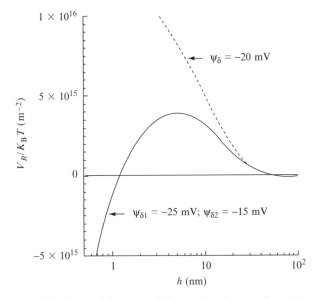

Figure 4.6. Potential energy of interaction for two flat plates.

close to its isoelectric point tending to stick to clay particle just as a result of electrostatic interactions. Of course, dispersion forces are also usually attractive and will aid the process.

4.3   Constant Charge or Constant Potential?

At this point, we should note that the charge at the Stern plane is related to the local potential gradient (Equation (4.15)). If we examine Figure 4.5, and keep the potential at the surfaces as fixed points, when we bring the two plates close together, the slope of the potential decreases due to the summation of the potential distributions. This means that the local charge changes (decreases) as a result of the constant potential. Hence, the results expressed by Equations (4.29) and (4.30) are strictly for that constant potential condition. In fact, we get an identical result with both the Debye–Hückel approximation and weak overlap condition if we keep the charge constant. At higher potentials and greater overlap, this is not the case and we have different results if we work at constant potential or constant charge. Where appropriate, the inclusion of ψ or σ in the subscript will be used to indicate if the equation relates to constant potential or constant charge. The choice of the appropriate equation would seem to be easily related to the type of surface that we have. For example, a surface with a charge due to isomorphous substitution in a lattice should be a good candidate for a constant-charge surface. On the other hand, a surface where the charge is controlled by the concentration of potential-determining ions in solution might best be treated as a constant-potential surface. In practice, the choice is rarely clear-cut. However, care should be exercised when using the constant-charge expressions at small separations as the approximations become inadequate, predicting much too strong a repulsion at $\kappa H < 1$.

5   THE INTERACTION BETWEEN TWO SPHERES

A similar approach to the repulsion between flat plates is taken to estimating the repulsion between two spheres. The problem of solving the potential distribution around a sphere is carried over to the problem of overlapping diffuse layers. However, there are a number of approximate expressions available (for example, see Hunter[1], Russel *et al.* [3] and Verwey and Overbeek [6]). Here, we give the approximate equations with an indication of where they can be used rather than discuss the details of the derivations.

5.1   Identical Spheres at $\kappa a < 5$

Here, the diffuse layer is becoming of a similar magnitude to the particle radius. With the Debye–Hückel approximation for the potential around each sphere, and simply summing the two, i.e. weak overlap, we have:

$$V_{\psi R} = 2\pi\varepsilon_r\varepsilon_0 a\psi_\delta^2 \exp(-\kappa H) \tag{4.32}$$

As with the case of flat plates, we have $V_{\sigma R} = V_{\psi R}$. This solution is best for weak overlap so that at $\kappa H > 2$ it is satisfactory over a wide range of κa. As $\kappa a \rightarrow 1$ or less, it is useful at close distances of approach [7].

5.2 Identical Spheres at $\kappa a > 10$

In this case, our diffuse layer is thin when compared to the particle radius. The results for low potentials, with h being the closest distance between the surfaces of the particles (i.e. $H = r - 2a$, with r the centre-to-centre distance), are:

$$V_{\psi R} = 2\pi\varepsilon_r\varepsilon_0 a\psi_\delta^2 \ln [1 + \exp(-\kappa H)] \tag{4.33a}$$

$$V_{\sigma R} = -2\pi\varepsilon_r\varepsilon_0 a\psi_\delta^2 \ln [1 - \exp(-\kappa H)] \tag{4.33b}$$

The constant potential expression, Equation (4.33a), works well for all separations and is acceptable down to $\kappa a > 2$ if at close approach such that $\kappa H < 2^3$. The constant charge expression, Equation (4.33b), should be used with caution, especially at close approach, as a large overestimate can be obtained.

5.3 Spheres of Radii a_1 and a_2 at $\kappa a > 10$

The equivalent form of Equations (4.33a) for spheres of any radii was given by Hogg *et al.* [8]:

$$V_{\psi R} = \frac{\pi\varepsilon_r\varepsilon_0 a_1 a_2}{a_1 + a_2}(\psi_{\delta 1}^2 + \psi_{\delta 2}^2)\left\{\frac{2\psi_{\delta 1}\psi_{\delta 2}}{(\psi_{\delta 1}^2 + \psi_{\delta 2}^2)}\ln\left[\frac{1 + \exp(-\kappa H)}{1 - \exp(-\kappa H)}\right]\right.$$
$$\left. + \ln[1 - \exp(-2\kappa H)]\right\} \tag{4.34}$$

5.4 A Sphere and a Plate

The result is that the potential energy is twice that estimated for two similar spheres, so that, for example, with $\kappa a < 5$ at weak overlap we can write:

$$V_{\psi R} = 4\pi\varepsilon_r\varepsilon_0 a\psi_\delta^2 \exp(-\kappa H) \tag{4.35}$$

and for close approach and large κa:

$$V_{\psi R} = 4\pi\varepsilon_r\varepsilon_0 a\psi_\delta^2 \ln [1 + \exp(-\kappa H)] \tag{4.36}$$

6 THE EFFECT OF PARTICLE CONCENTRATION

Up to this point, we have only considered the interaction of two isolated colloidal particles. It is reasonable to neglect the presence of other particles if the time-average separation of the particles is very much larger than the range of the diffuse layer. Once the separation becomes similar in magnitude to the range of the diffuse layer, it becomes untenable to consider the osmotic pressure between a pair of particles relative to a 'bulk value'. The background electrolyte now contains other particles with their counter-ions. The simplest approach is to consider the additional concentration of ions due to the particle counter-ions as an addition to the ions of the background electrolyte. Of course, each charged particle is a 'macro-ion' but the number is very much smaller than the number of the corresponding counter-ions, and hence the particle contribution may be ignored without introducing a large error. However, once the volume fraction of the particles becomes high, the volume occupied by the particles must be allowed for as this volume is excluded to the ions. Russel *et al.* [3] give us a convenient expression for κ for a symmetrical electrolyte with ions of valency z as follows:

$$\kappa = \left\{ \left(\frac{e^2}{\varepsilon_r \varepsilon_0 k_B T} \right) \frac{\left[2z^2 n_0 - \left(\frac{3\sigma_0 z\varphi}{ae} \right) \right]}{(1 - \varphi)} \right\}^{\frac{1}{2}} \tag{4.37}$$

The $(1 - \varphi)$ term in the denominator corrects the ion concentration for the volume occupied by particles, while σ_0 is the surface charge density which is often measurable by titration. If the Stern layer charge was available, it would be better to use that value as the strongly bound counter-ions are effectively removed from osmotic activity. The ζ-potential could be used to give an estimate of σ_δ; $(\sigma_0/e)(4\pi a^2)$ is the number of charges per particle of radius a and surface charge density of σ_0, and so $(\sigma_0/ze)(4\pi a^2)$ is the number of counter-ions associated with each particle, while $3\varphi/4\pi a^3$ is the number of particles per unit volume. The product of these last two terms and z^2 gives the counter-ion contribution to the equation. Now, the counter-ions are assumed to have the same activity as would an ion in bulk solution and so we are not making any correction for the structuring effect of the charged particles. To do so would require a complex statistical mechanical analysis. However, the approximation given in Equation (4.37) is a useful correction for many concentrated dispersions. This is most important for small charged particles where we have added little extra electrolyte.

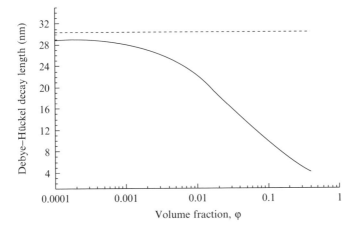

Figure 4.7. The Debye–Hückel parameter as a function of φ: $a = 200\,nm$; [NaCl] $=$ $10^{-4}\,mol\,dm^{-3}$; $\sigma_0 = 10\,\mu C\,cm^{-2}$. The dashed line represents the limiting value for zero volume fraction.

As an illustration of the problem, we may consider two suspensions of charged spherical particles in a dilute solution of sodium chloride. Polymer latex particles would be a good example here as they can be produced with quite high surface charge densities and also in many different sizes. Figure 4.7 shows how the range of the diffuse layer can vary with particle concentration with a plot of the Debye–Hückel decay length, κ^{-1}, against the volume fraction, φ. There is nearly a factor of 10 change in the Debye length which, because of the square-root relationship with ion concentration, means that the background electrolyte is completely swamped by the counter-ion concentration as the dispersion becomes concentrated. At volume fractions of ~ 0.35 we are at similar concentrations to those found in many coatings. The effect on the electrostatic repulsive potential is illustrated in Figure 4.8 where curves are plotted for three concentrations of particles. The value of $10\,\mu C\,cm^{-2}$ does represent a higher value than those found in many systems, which are often in the range 0.1 to $1\,\mu C\,cm^{-2}$. However, even at these lower values we obtain values of $12\,nm < \kappa^{-1} < 20\,nm$ at high solids contents, and so the exponential functionality in the repulsive energy expression is still a significant correction.

The addition of ionic surfactants to dispersions is common. They are added to aid dispersion and to provide charge to prevent aggregation. The density of surfactant molecules on the surface of the particles usually corresponds to a monolayer and so the surface charge density is high. However, the Stern

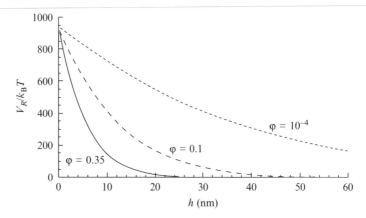

Figure 4.8. Electrostatic repulsion potential for three different concentrations of particles (using Equation (4.33a)): $a = 200\,\text{nm}$; $[\text{NaCl}] = 10^{-3}\,\text{mol}\,\text{dm}^{-3}$; $\psi_\delta = 80\,\text{mV}$; $\sigma_\delta = 10\,\mu\text{C}\,\text{cm}^{-2}$.

charge is lower as many counter-ions are strongly bound but it is also important to recognize that the ionic surfactants are also electrolytes. The problem is a little more complex though as the surfactant molecules form micelles which limit their numbers in free solution and the micelles also bind the counter-ions to the amphiphilic molecule. However, the diffuse layer counterions should still be included and the micelles can act as small particles and we will need to keep this in mind when we consider the behaviour of concentrated dispersions.

7 STERIC INTERACTIONS

In many colloidal systems we add non-ionic materials which adsorb onto the surfaces of our particles. The purpose is to prevent aggregation by keeping the surfaces apart. For this to be effective, the molecules must be firmly attached but must also extend into the solution so that the surfaces cannot approach too closely, and this gives the term 'steric interaction'. The types of molecules that are most effective for this purpose are non-ionic surfactant block copolymers. (Charged species such as proteins and polyelectrolytes are also used but we will delay discussion of these until we discuss the full picture of colloid stability.) Simple homopolymers are rarely used for this purpose as they need to be chemically grafted to the surface to prevent desorption if they are lyophilic, and if they are lyophobic they will be difficult to get into solution and would form a dense ineffective layer. In this section, we will concentrate on materials which are effective stabilizers as this is where the important applications lie.

So, the picture emerging is of molecules with one or more strong anchoring moieties but with other moieties which are in a good solvent. This means that we can design effective systems which can operate as stabilizers in high electrolyte concentrations or in non-aqueous (or better, low polarity) media. The simple linear non-ionic surfactants can be used as well as larger molecules. The model that we are going to work with is relatively simple but it is important to recognize that the problem is many faceted and necessarily complex if a full treatment is to be attempted. In outline, what we will discuss is the approach of two surfaces which are fully covered by strongly attached polymer (or non-ionic surfactant) layers. As the surfaces approach, the outermost parts of the layers start to mix as the molecules interpenetrate. The result is an increase in free energy. At close approach, we may picture an additional constraint as the surface–surface separation becomes less than the dimension of a single layer. This would result in a further increase in free energy as the chains are 'compressed', that is, there are even fewer possible configurations available to each molecule, thus resulting in a marked decrease in entropy. However, this is not a problem that we shall have to deal with when considering good steric stabilizers as we shall see that the repulsion becomes very large before such close approach is attained. Thus only the outermost parts of the layers in systems are of practical importance although we could construct systems where this would not be the case.

Before we can begin to construct our model, we must have some information on the composition of the layers attached to the surfaces. Figure 4.9 illustrates some of the scenarios that we must consider. The schematic shown in Figure 4.9(a) illustrates a surface covered with terminally anchored polymer molecules. The local polymer concentration, ρ_2, shows a maximum a little way away from the surface with the concentration falling further out. Remember that the coils are dynamic and we are trying to obtain the closest lateral spacing. The profile is distorted from that which we could expect from an isolated coil, first, by the proximity of the particle surface, and secondly, by the crowding of the neighbouring coils. Hence, the extension away from the surface should be a little greater than the dimension of an isolated coil in solution. If we consider a block or graft copolymer strongly attached to the surface, we may expect the profile to be a little denser as the adsorbing sections of the molecule force the soluble moieties closer together. The extension away from the surface is a maximum for the longest tails where only one end is attached. Loops are pulled to the surface at both ends and tend to have a shorter extension. The adsorbed sections ('trains'), of course, are held close to the surface. This is illustrated in Figure 4.9(b) where the profile is shown as relatively uniform until the outer portion of the tails are reached. Lastly, in Figure 4.9(c) we illustrate an adsorbed surfactant layer which has a uniform concentration profile with a short-range reducing profile due to the variation in chain length. The polydispersity of most surfactants tends to be less than

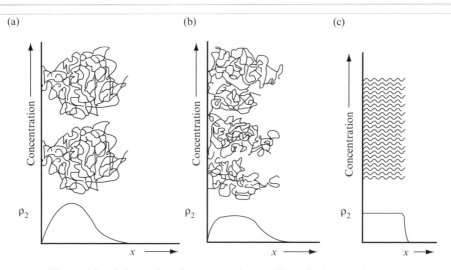

Figure 4.9. Schematics of concentration profiles of adsorbed layers.

we find with most polymer systems. The polydispersity of the latter means that the outer profile of the layer can be expected to fall more slowly than we might expect from the periphery of an isolated chain. In summary, we would like to have the following three pieces of information:

- the amount of material attached to a unit area of the surface
- an adsorbed layer 'thickness'
- the profile of the outer part of the layer

The experimental estimation of each of these features has been discussed earlier in Chapter 2. In this section, we will restrict our discussion to the interaction between two coated particles. Furthermore, the model will be restricted to that of layers of uniform concentration as this will simplify the algebraic description and make the route more transparent. This represent a less serious limitation than might be apparent at first sight as effective steric stabilization is frequently provided by densely packed layers of non-ionic surfactants or branched polymer chains forming part of a block copolymer. The interaction calculated by using this model will result in a steeper pair-potential than may actually be the case, but the softness is nearly always the result of stabilizer polydispersity and this is not easy to model effectively. We will return to this point at the end of the discussion.

The starting point for our model is the recognition that the steric inter-action is zero until the outer periphery of the polymer layer on one particle comes into contact with that of an adjacent particle. Closer approach means that the polymer or surfactant chains start to mix. The chain concentration in the overlap volume is higher than that in either of the isolated layers and this results in an osmotic pressure acting to separate the particles. If the thickness of the layer is δ, the interaction occurs when the particle surfaces are at separations $< 2\delta$. This is the interaction that we will model.

Should the particle separation become $< \delta$, a second effect should be modelled. The surface of the second particle further restricts the possible conformation of the chains attached to the opposing surface, giving rise to an additional contribution to the interaction energy. This *volume restriction* term will be neglected here as in any successful steric stabilizer this degree of the overlap of layers will never occur as only a small degree of overlap is sufficient to produce potentials much greater than $k_B T$.

In Figure 4.10, we see the overlap of adsorbed layers of uniform concentration for several geometries for layers which have a thickness of δ and with a particle surface-to-surface separation distance of h. The concentration of stabilizer in each layer is c_2 and so the concentration in the overlap region is $2c_2$. This means there is a higher osmotic pressure in the region where the two layers are mixing than before they came into contact. We will define this as the excess osmotic pressure, Π_E, which acts to separate the particles. In order to calculate the force, we must define the area over which the pressure is acting. For the examples illustrated in Figure 4.10, this is the unit area for the plates and the area at the base of the spherical cap which is located at the mid-plane. The interaction energy between the particles (which originates from the free energy of mixing the polymers in the solvent) is then the difference in the free energy of two particles at overlap relative to the value just at contact. This value can be calculated by integrating the force with the distance between the particles:

$$V_S = \int_{2\delta}^{h} -\Pi_E A \mathrm{d}x \tag{4.38}$$

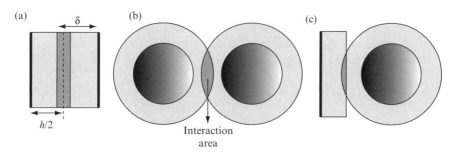

Figure 4.10. Schematics of steric interactions – overlap of adsorbed layers for different geometries: (a) plate–plate; (b) sphere–sphere; (c) sphere–plate.

Of course, the integral of '$A dx$' gives us the overlap volume which we may define as v_o.

The Flory–Huggins model of polymers in solution gives us the osmotic pressure of a polymer solution as a function of concentration (as discussed earlier in Chapter 2):

$$\Pi = \left(\frac{RTc_2}{M_2} + RTB_2c_2^2 + \ldots \right) \tag{4.39}$$

where B_2 is the second virial coefficient:

$$B_2 = \left(\frac{\bar{v}_2}{M_2} \right)^2 \frac{1}{\bar{v}_1}(0.5 - \chi) \tag{4.40}$$

with the subscripts '1' and '2' referring to the solvent and polymer species, respectively, as earlier. We can now use Equation (4.39) to calculate the energy difference as:

$$V_S = -(2v_o\Pi_{x = 2\delta} - v_o\Pi_{x = h}) \tag{4.41}$$

and so:

$$V_S = \left\{ \left[\frac{RT2c_2}{M_2} + RTB_2(2c_2)^2 \right] - 2\left(\frac{RTc_2}{M_2} + RTB_2c_2^2 \right) \right\}v_o \tag{4.42}$$

which gives us:

$$V_S = 2RTB_2c_2^2v_o \tag{4.43a}$$

$$V_S = \frac{2RT}{\bar{v}_1} \left(\frac{c_2\bar{v}_2}{M_2} \right)^2 (0.5 - \chi)v_o \tag{4.43b}$$

We should note in Equation (4.43) that the term, which is squared in parentheses, is just the volume fraction of polymer in the layer. For the three geometries illustrated in Figure 4.10, we have the overlap volumes as follows:

$$\text{plate} - \text{plate}, \ v_o = 2\delta - H \tag{4.44a}$$

$$\text{sphere} - \text{sphere}, \ v_o = \frac{2\pi}{3}(\delta - H/2)^2(3a + 2\delta + H/2) \tag{4.44b}$$

$$\text{sphere} - \text{plate}, \ v_o = \frac{4\pi}{3}(\delta - H/2)^2(3a + \delta + H) \tag{4.44c}$$

There are the same expressions that are given by Hunter [1] and Napper [9]. Other models are also available, e.g. Russel *et al.* [3], Dolan and Edwards [10] and DeGennes [11], but the above are sufficient for our purpose here.

The expressions given in Equations (4.43a) and (4.43b) indicate what we can expect for sterically stabilized particles. The major features are:

(a) There is a marked change with temperature.
(b) The repulsion increases rapidly as the polymer concentration in the layer is increased.
(c) There is an increase in repulsion as the separation decreases.

An increase in temperature can be anticipated to give an increased repulsion with non-aqueous systems directly from the 'RT' term, but also from the decrease in the χ-parameter as the solubility improves. However, especially in aqueous systems, this is not always the case as the $\chi \to 0.5$ as $T \to$ the LCST (where LCST is the *lower consolute solution temperature*, above which phase-separation of the polymer takes place). As this occurs, the repulsion approaches zero, the adsorbed layer collapses, and further temperature increases result in attraction.

As shown in Equations (4.43a) and (4.43b), the repulsion increases with the square of the polymer concentration in the adsorbed layer. A more rapid increase than this would have been included if we had not restricted ourselves to the simple Flory–Huggins treatment. In principle, we could use the osmotic pressure obtained experimentally from polymer solutions of similar concentration. However, that would not necessarily provide a better description as polymer attached to a surface is not quite the same as similar molecules that are free in solution. In addition, the model assumes that a higher polymer concentration is localized to the overlap volume and this is an oversimplification. Some rearrangement must occur, resulting in a larger but poorly defined interaction volume. This 'local dilution' would easily counteract the improvement of a more complex expression for the osmotic pressure due to the latter's marked dependence on concentration. At this point, it is relevant to recall the approximations made in the Flory–Huggins model. This was based on a simple mixing concept, using a lattice model with the molar volumes of the solvent and the chain segments being equal. The χ-parameter was expressed as an enthalpic term. Now, the latter depends on both the temperature and the polymer concentration as there is also an entropic contribution [9]. The value of χ increases as both the temperature and polymer concentration increase. This would reduce the magnitude of the interaction energy as the surfaces approach.

The biggest problem, however, is concentration detail in the early overlap regime. This is the point at which polydispersity is really controlling the softness of the interaction and a detailed profile of the outermost layer

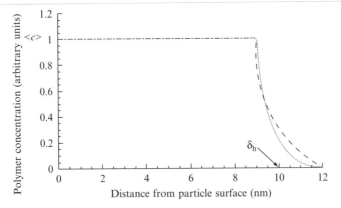

Figure 4.11. Illustration of the simplified model for adsorbed polymer layers, using the relationships shown in Equations (4.45a–4.45c) to give the concentration decrease of the outer layer: ($\cdots\cdots$) polymer concentration, exponential fall (Equation (4.45b)); (– – –) polymer concentration, cube root fall (Equation (4.45c)).

concentration is needed. Although there are some models for the concentration profiles of adsorbed layers and some experimental measurements, such information is poorest at the outermost extremities of the layers. This is the region that is of the greatest interest in systems with good steric stabilizers as it is the curvature here that we require when we need to estimate interparticle forces or rheological properties.

In order to illustrate how the system could be modelled, let us consider the simplified model for the concentration profiles shown in Figure 4.11. Here, there is a uniform polymer concentration extending from the surface to a distance δ_1. The polymer concentration then falls monotonically until it is negligible, at δ_m. A practical system which would have a profile somewhat similar to this would be a particle stabilized with poly(12-hydroxystearic acid), a useful stabilizing moiety for particles dispersed in hydrocarbons. This heavily branched stabilizer is predominantly a hexamer but with pentamers and septamers also present. The profiles shown in Figure 4.11 follow the following equations:

$$x < \delta_1, c = <c> \tag{4.45a}$$

$$\delta_1 < x < \delta_m, c = \exp\left[-5\left(\frac{x - \delta_m}{\delta_m - \delta_1}\right)\right] \tag{4.45b}$$

or:

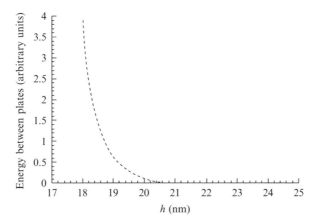

Figure 4.12. Illustration of the interaction between two plates with polymer 'brush' layers, displaying an exponentially decaying concentration at the outer periphery of each layer.

$$c = 1 - \left(\frac{x - \delta_m}{\delta_m - \delta_1}\right)^{\frac{1}{3}} \tag{4.45c}$$

It should be stressed that both rates of decay are purely illustrative. The exponential decay rate was chosen to give a very low concentration at δ_m. A hydrodynamic thickness of $\delta_h = 10\,\text{nm}$ is also identified on this figure and serves to illustrate that we might expect repulsion prior to a surface–surface separation of $2\delta_h$. We can use the relationships shown in Equations (4.45a–4.45c) in conjunction with those in Equations (4.44a–4.44c) to describe the concentration variation prior to the integration to give the energy. Figure 4.12 illustrates the results obtained for two plates covered with polymer 'brushes' whose concentration profile could be that shown as the exponential decay in Figure 4.11. It is interesting to note that not only is the decay very rapid as the interaction 'follows' the square of the concentration but that we do not see a significant repulsion until the surface–surface separation approaches $2\delta_h$. After this, it rises rapidly over the next 1–2 nm.

8 CALCULATION STRATEGY

Although the above discussion illustrates that a knowledge of the details of the concentration profile is desirable, it is difficult to obtain this information experimentally. Techniques such as neutron scattering loose their sensitivity at the low local concentrations at the periphery. In many practical circumstances, we prepare colloidal systems with the most effective steric stabilizers that we can. Thus, we can frequently use a uniform densely packed profile and the value of δ_h can be measured. The interaction energy will increase very rapidly as soon as the two layers come into contact. This situation can be approximated to:

$$H > 2\delta_h, \quad V_S = 0 \qquad (4.46a)$$

$$H < 2\delta_h, \quad V_S = \infty \qquad (4.46b)$$

In the case of non-ionic surfactants, the stretched surfactant chain length would give a suitable distance for the repulsive wall. When combining this with the attractive potentials, we can get an indication as to whether or not the attraction has been reduced to a few units of '$k_B T$' or less so that the system can readily be handled. Only occasionally will we require more detailed information of the potential and then we have to model the concentration profile of the stabilizer layer. However, in aqueous systems, or those of intermediate polarity, the electrostatic repulsion will control the pair-potential profile; the ζ-potential is the potential at a distance from the surface corresponding to δ_h.

As an illustration, we may take the problem of the interaction of terminally anchored polymer chains which has been analysed in some detail by Dolan and Edwards [10] and deGennes [12]. The expression using the latter's work for the force between two planar surfaces is as follows:

$$F_T(h) = k_B T \Gamma^{3/2} \left[\left(\frac{H}{2\delta}\right)^{-9/4} - \left(\frac{H}{2\delta}\right)^{3/4} \right] - \frac{A}{6\pi H^3} \qquad (4.47)$$

where Γ is the number of chains per unit area of surface. As an illustrative calculation, we may take a δ value of 10 nm, a surface density of polymer of one chain every 100 nm^2 and a Hamaker constant for the system of $A \sim 10^{-20}$ J. The results obtained are presented in Figure 4.13 and show the very rapid rise in repulsion at 2δ as the two 'brush' layers begin to overlap. Of course, the polydispersity is not taken into account. The very steep wall is of less interest than the details of the minimum at the start of the near vertical rise, and this is where chain polydispersity and weak electrostatics are likely to control the profile.

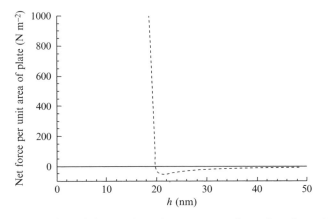

Figure 4.13. Illustration of the net force between two plates: δ and $s = 10\,\text{nm}$; $A = 1 \times 10^{-20}\,\text{J}$.

REFERENCES

1. R. J. Hunter, *Foundations of Colloid Science*, Vol. I, Oxford University Press, Oxford, UK, 1987.
2. J. Lyklema, *Fundamentals of Interface and Colloid Science*, Vol. 1, Academic Press, London, 1991.
3. W. B. Russel, D. A. Saville and W. R. Schowalter, *Colloidal Dispersions*, Cambridge University Press, Cambridge, UK, 1989.
4. A. L. Loeb, J. Th. G. Overbeek and P. H. Weirsema, *The Electrical Double-Layer Around a Spherical Colloid Particle*, MIT Press, Cambridge, MA, USA, 1961.
5. H. Ohshima, T. W. Healy and L. R. White, *J. Colloid Interface Sci.*, **89**, 446 (1982).
6. E. J. W. Verwey and J. Th. G. Overbeek, *Theory of Stability of Lyophobic Colloids*, Elsevier, Amsterdam, 1948.
7. A. B. Glendinning and W. B. Russel, *J. Colloid Interface Sci.*, **93**, 95 (1983).
8. R. Hogg, T. W. Healy and D. W. Fursteneau, *Trans. Faraday Soc.*, **62**, 1638 (1966).
9. D. H. Napper, *Polymeric Stabilisation of Colloidal Dispersions*, Academic Press, London, 1983.
10. A. K. Dolan and S. F. Edwards, *Proc. R. Soc. London, A*, **337**, 509 (1974).
11. P.-G. de Gennes, *Adv. Colloid Interface Sci.*, **27**, 189 (1987).
12. P.-G. de Gennes, *Macromolecules*, **15**, 492 (1982).

Chapter 5

The Stability of Dispersions

1 INTRODUCTION

The term 'stability' has two distinct meanings in the context of colloidal dispersions. In many practical situations, it is taken to mean that there are no signs of phase-separation over a period of time. If particles in a dispersion showed a tendency to sediment or cream over a period of storage, that dispersion would be termed 'unstable'. However, we also use the term in another context to mean that the particles have no tendency to aggregate. With large or dense particles, these two usages can be contradictory. With very small particles, this may not be the case. Hence, we will refer to *colloid stability* when we mean that the particles do not aggregate and *mechanical stability* when they do not sediment. The key to understanding colloid stability is the pair-potential. In Chapters 3 and 4, we have developed expressions describing the distance-dependence of the various components of the potential energy of interaction of two particles. The total potential energy of interaction, i.e. the *pair-potential*, is calculated as the sum of each of these components.

2 THE STABILITY OF CHARGE-STABILIZED COLLOIDS – THE DLVO THEORY

The linear combination of the dispersion force contribution to the pair-potential with the electrostatic repulsion gave the first comprehensive model framework for the stability of colloidal dispersions. This work was due to Derjaguin and Landau[1], and independently by Verwey and Overbeek[2] – hence the term, *DLVO theory* – and we may summarize the total interaction energy as follows:

Colloids and Interfaces with Surfactants and Polymers – An Introduction   J. W. Goodwin
© 2004 John Wiley & Sons, Ltd   ISBN: 0-470-84142-7 (HB)   ISBN: 0-470-84143-5 (PB)

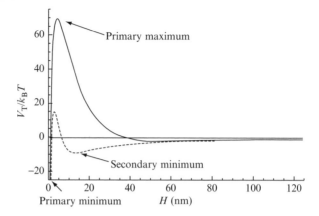

Figure 5.1. The pair-potential calculated for the rutile form of titanium dioxide at concentrations of 1 mM (——) and 10 mM (---) of sodium chloride.

$$V_T = V_A + V_R \tag{5.1}$$

So, for example, we may use Equation (3.25b) for V_A and Equation (4.33a) for V_R (see earlier) to describe the interaction between two titanium dioxide particles as a function of separation at different electrolyte concentrations. The results of such a calculation are illustrated in Figure 5.1. for 100 nm radius rutile particles with a ζ-potential of -45 mV. The calculation is for an isolated pair of particles and so we are considering very dilute dispersions. As V_A and V_R have a quite different dependence on particle separation, the combined curve has quite a complicated structure. The general features are as follows:

(1) A *primary minimum*, V_{pmin}, where the dispersion interaction is much larger than the electrostatic term. The details here are uncertain, however, as details of the molecular structures of both the surface and the adsorbed ions are of a similar scale to the separation.
(2) The *primary maximum*, V_{max}, which occurs at a distance a little further away where the electrostatic interaction is dominant.
(3) After this, the interaction energy decreases to the *secondary minimum*, V_{smin}, as the attractive interaction again becomes a little larger than the repulsion.

The primary minimum indicates that the aggregated state is the lowest-energy condition and this is where we would expect the particles to reside. The primary maximum opposes the close approach of the particles and is an activation energy that must be exceeded for aggregation to occur. The motion of the particles is governed by the thermal energy and, of course, we may describe the energy distribution by the Boltzmann equation. This is important as we can see that we are predicting a *kinetic* stability as the rate of aggregation should be proportional to exp $(-V_T/k_B T)$. When $V_T \gg k_B T$, our particles will be in a colloidally stable state. The value of $V_{smin} \sim k_B T$ in our example and thus we may expect many of the particles to be in close proximity for much of the time. In other words, as the net attractive energy is only a little larger than the average thermal energy, any aggregation in the secondary minimum is reversible. We will return to discuss the kinetics of the aggregation of particles later in this chapter.

In Chapters 3 and 4, we discussed the various contributions in some detail. If there is a net repulsive interaction between two particles such that $V_T \gg k_B T$, the particles will not aggregate. If there is a strong attraction so that $-V_T \gg k_B T$, we will get strong aggregation or *coagulation* of the particles with the particle surfaces coming into close proximity. In the case of fluid particles, this is the precursor to coalescence. There are other situations where the particles are attracted at long distances but do not come into close contact. Here, we will use the term *flocculation* to describe the aggregation. Many authors use the terms interchangeably but it is useful to separate them. For example, particles can be flocculated in a shallow attractive well but still have a large energy barrier to overcome before they can come into close contact. Some examples here are aggregation in the secondary minimum predicted for some electrostatically stabilized dispersions, or aggregation due to the depletion forces produced when a non-adsorbing polymer is added to a dispersion. In both of these situations, the well is $-1 > V_{min}/k_B T > 100$ and the aggregates may be redispersed with moderate shearing forces, unlike the coagulated state. The term *bridging flocculation* is used to describe the aggregation of particles by the adsorption of polymer on two particles simultaneously, thereby 'tying them' together.

3 MECHANISMS OF AGGREGATION

3.1 Electrostatically Stabilized Dispersions

One of the great early successes of the DLVO theory of formulation of the pair potential for electrostatically stabilized particles was the ability to predict their coagulation on the addition of electrolytes [3]. The addition of an indifferent electrolyte reduces the range of the repulsive component and the maximum in the potential energy curve is reduced. An indifferent electrolyte is one which does not contain a potential-determining ion, such as Ag^+ with silver halide particles or H_3O^+ with oxide particles. At some point, the value of the maximum approaches zero and there is no barrier to particles coming into close contact due to the dispersion forces. Of course, the concentration and type of the counter-ions does have an effect other than compression of the diffuse layer. The adsorption in the Stern layer can also change which we can readily observe experimentally via changes in the ζ-potential. The latter is frequently assumed to equate to the Stern potential, ψ_δ, and is therefore used in calculations of the pair-potential. The variation in the tendency to adsorb in a series of ions of the same valency is known as the *Hofmeister series* and these ions' specificities also show up in other colloidal features such as micellization [3]. The series for monovalent cations and anions is as follows [2]:

$$Cs^+ > Rb^+ > K^+ > Na^+ > Li^+$$

$$CNS^- > I^- > Br^- > Cl^- > F^- > NO_3^- > ClO_4^-$$

With cations, the small size of the lithium ion results in a higher charge density which, in turn, means that it is more strongly hydrated. The larger anions are more easily polarized and this increases the adsorption. We see a stronger effect of the ion type with positive particles and anionic counter-ions than with negative particles and cationic counter-ions.

The valency of the counter-ion is extremely important. The coagulating power of an ion increases dramatically with its valency, as encompassed in the Shultz–Hardy rule [3–5] which states that the coagulating power varies as z^6 where z is the counter-ion valency. Experimentally, this is not always the case and a lower exponent can be found. Figure 5.2 illustrates the conditions for

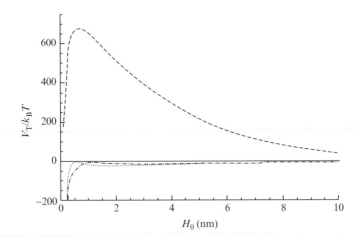

Figure 5.2. The pair-potential calculated for polystyrene latex particles with a radius of 500 nm as a function of the ζ-potential and sodium chloride concentration: (—·—·) −20 mV, 100 mM; (·······) −30 mV, 400 mM; (- - - -) −50 mV, 10 mM.

the loss of stability. The data shown in this figure were obtained for polystyrene latex particles with a diameter of 1 μm and were calculated for different combinations of ζ-potential and electrolyte concentration resulting in the conditions for instability being met. The unstable particles are those where the conditions pertain that:

$$V_{max} = 0, \text{ and so at this point } \frac{V_R}{V_A} = -1 \tag{5.2a}$$

and also:

$$\frac{\partial V_T}{\partial H} = 0 = \frac{\partial V_R}{\partial H} + \frac{\partial V_A}{\partial H}, \text{ with } \kappa \to \kappa_{ccc} \tag{5.2b}$$

If we take the equations for two similar flat plates, i.e. Equations (3.21) and (4.30) (see earlier), and differentiate these, we have:

$$0 = -\kappa_{ccc} V_R - \frac{2}{H} V_A \tag{5.3a}$$

or:

$$\kappa_{ccc} H = 2 \tag{5.3b}$$

and so putting this value back into the expression for V_{max}, we have, from Equations (3.21) and (4.30):

$$\frac{64 n k_B T}{\kappa_{ccc}} \tanh \left(\frac{z e \psi_\delta}{k_B T} \right) \exp(-2) = \frac{A_{121} \kappa_{ccc}^2}{48 \pi} \tag{5.4}$$

The expression for κ was given earlier by Equation (4.10a). Substitution of this into Equation (5.4) gives us the following result for the critical coagulation concentration (ccc):

$$ccc \propto \frac{1}{z^6} \tag{5.5}$$

However, what often occurs as the electrolyte concentration is increased is that the potential, ψ_δ, falls to a low value and then we find experimentally that we have the weaker dependence:

$$ccc \propto \frac{1}{z^2} \tag{5.6}$$

What we observe experimentally is that as we approach the critical coagulation concentration (ccc), we start to see aggregates in suspension. Hence, we can determine the value by simply using a series of test tubes containing a range of electrolyte concentrations, to each of which we add a known amount of dispersion. After a few minutes, aggregates will be seen at concentrations above the ccc. This experiment can be made a little more precise if a spectrophotometric determination of turbidity is carried out. The procedure here is as follows:

(1) Mix a known volume of the dispersion with known volumes of different electrolyte concentrations.
(2) Allow these to stand for a fixed time – 30 min would be suitable.
(3) Centrifuge the dispersions at a low *g*-value so that only the aggregates are removed.
(4) Measure the turbidities of the supernatants.

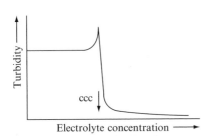

Figure 5.3. The turbidity of the supernatant from a dispersion as a function of concentration of added electrolyte.

A curve similar to that shown in Figure 5.3 will be obtained.

It is important to be aware that the coagulating powers of multivalent cations can be a little more complex than we might expect initially. Trivalent cations such as Al^{3+} and La^{3+} only exist in this form at low pH values. As Matijévic has pointed out [6], as the pH approaches values of 7 and above, these ions exist as large complex highly charged species. Hence, we should begin to think of these counter-ions as small counter-charged nanoparticles which can cause hetero-coagulation, and this is why they are such effective coagulants.

4   HETERO-COAGULATION AND HETERO-FLOCCULATION

This is an important aspect of the aggregation of colloidal particles in many practical situations. We will discuss this in broader terms here than is often done. Thus, under this heading we are including aspects of the interaction of particles which can carry an opposite charge from each other and the effects of polyelectrolytes with a charge different in sign from the particles. It was pointed out in the previous paragraph that ions such as Al^{3+} only exist in this form at values of pH $<\sim$ 4. At higher pH values, large hydrolysed complex ions are formed with high charge densities. These can be very strongly adsorbed to the surfaces of negatively charged particles. Clearly, if present in high concentration, the surface charge can be reversed in sign in exactly the same manner as occurs with high concentrations of the simple multivalent ions and the system would be potentially 're-stabilized'. Hence, when used as coagulants, the concentrations are kept low. Neutralization of the surface charge is one mechanism for aggregation just as it is with simple multivalent ion species. The diffuse layer compression is, of course, occurring simultaneously, as discussed in the previous section, but the strong binding has a very large effect in the reduction of the Stern potential, ψ_δ. However, complete neutralization of the charge is not a prerequisite for aggregation to occur. There are two further possibilities. First, the small macro-ions can act as bridges at low concentrations. Secondly, the adsorption can produce 'patches' on the surface of different charge so that on close approach local attraction can occur. This is just 'charge patch' flocculation, as suggested by Gregory [7], as a mechanism for the aggregation of negatively charged particles by cationic polymers. For this to work, the background electrolyte concentration must be sufficiently high that an approaching particle surface can 'recognize' such a local difference with the resulting attraction. This means that the patches have to be of a similar magnitude to the decay distance of the diffuse layer potential, κ^{-1}.

4.1 Polymeric Flocculants

Cationic polyelectrolytes, such as a positively charged polyacrylamide-based copolymer, can form bridges, causing hetero-flocculation. The radius of gyration of a polyelectrolyte is much larger than if the polymer were uncharged as the charged groups repel each other, thus forcing the coil to take up an expanded configuration. The dimension of the coil is a function of electrolyte concentration as the screening of the charges varies, and in many cases also the pH. Bridging by polymers (or by other particles) is most effective if the particles are small and concentrated enough that their average separation is of a similar magnitude to the maximum dimension of the coagulant species. This will not be the case with larger colloidal particles and then the charge patch mechanism becomes the likely route. The critical factor here is that the coagulating polymer species will have sufficient time to be adsorbed 'flat' on the surface before an encounter with another particle can occur. A schematic of this situation is shown in Figure 5.4. Both bridging and

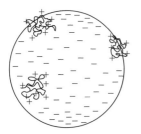

Figure 5.4. Schematic of 'charge patch' formation by the adsorption of a cationic polyelectrolyte on an anionic particle.

charge patch flocculation work well when the surface coverage is less than 50 % (by adsorbed polymer). As greater coverage is attained, there is an increasing tendency for a stable system to be engineered. Polymeric flocculants are normally used at concentrations of the order of parts per million (ppm).

Anionic polyelectrolytes, such as poly(acrylic acid) and partially hydrolysed polyacrylamide, can also be utilized as polymeric flocculants. However, many colloidal systems of practical importance are anionic also. In this case, divalent ions such as Ca^{2+} can be added and act as ion bridges to bind the polymer to the surfaces and promote bridging flocculation. Calcium ions bind very strongly to carboxyl groups and this can help to attach the anionic polymer to material coated with proteinaceous material. Ion bridges between two separately coated particles can also be formed, hence producing a flocculated system.

Depletion flocculation should also be considered at this point of the discussion. This mechanism though is not useful in terms of separation processes as the attractive well is too shallow. The flocculated state is readily re-dispersed and so the particles could not be successfully separated, for example, by pressure filtration. In fact, this reversibility is a feature that can be used to advantage in some situations as we shall see when we discuss mechanical stability.

4.2   Particulate Hetero-coagulation

The charge patch flocculation mode described above is just one example of hetero-coagulation where surfaces with charges of different sign attract one another. The dispersion force contribution is also present so that the total attractive force is large. An example of hetero-coagulation, which is of immense practical importance, is to be found in the colloidal behaviour of clays. Kaolinite is an aluminosilicate which has a 1:1 layer lattice structure. The clay crystal is made up of alternating layers of silica tetrahedra and alumina octahedra connected via shared oxygen atoms. The basal planes are silica and the crystals are irregular hexagons which have an axial ratio of approximately 10:1. As with all oxides, the silica surfaces are amphoteric so that the surface hydroxl groups can either ionize or bind a proton:

$$O-\underset{\underset{O}{|}}{\overset{\overset{O}{\diagdown}}{Si}}-O^{-} \rightleftharpoons O-\underset{\underset{O}{|}}{\overset{\overset{O}{\diagdown}}{Si}}-O-H \rightleftharpoons O-\underset{\underset{O}{|}}{\overset{\overset{O}{\diagdown}}{Si}}-\overset{+}{O}\underset{\diagdown H}{\overset{\diagup H^{-}}{}}$$

The point of zero charge (pzc) occurs at a low pH for silica and hence the surfaces would be negatively charged. However, there is an additional factor in that in the formation of the clay, some of the silicon atoms were replaced

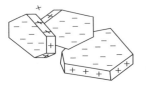

Figure 5.5. Schematic representation of 'edge–face' coagulation found with kaolinite at pH values < 7 when the edges carry a positive charge and the faces are negative. The particles form a very open structure, referred to as 'card-house' aggregation structure.

by aluminium, and some of the aluminium in the next layer down by magnesium. The crystal structure, although distorted, is basically the same as the cations are smaller than the oxygen anions. This is known as *isomorphous substitution* and the result is that there is an excess negative charge due to the replacement of ions by some of lower valency. This charge is balanced by positive ions in the Stern and diffuse layers. The surfaces are therefore strongly negatively charged. Isomorphous substitution is only present in the outermost crystal layers and is probably one of the factors that limits the growth. The crystal edges have exposed alumina layers and the pzc of this edge surface is close to that of alumina. The result is that the edges are positive at pH < 7 and become increasingly negative at pH > 7. Hence, at pH values < 7 the clay particles aggregate in an edge–face structure similar to a 'house of cards'. Figure 5.5 illustrates this mode of hetero-coagulation.

Another situation in which hetero-coagulation can occur is in mixed particle systems, especially when at least one of the components is an oxide or a particle coated with a layer which has an isoelectric point at a different pH from the other component. In this situation, the van der Waals attraction will be aided by the attraction of oppositely charged surfaces over some of the pH range. The readjustment of the pH of the dispersion after aggregation has taken place will not result in re-stabilization of the aggregated particles due to the large amount of energy required to overcome the deep primary minimum attractive energy. The exception would be particles coated with a protein layer. In this case, the expansion of the hydrophilic layer due to the internal charge repulsion may be sufficient to aid the 're-peptization'.

The mixing of particle systems of different surface charges is important. To illustrate this, we can take the example of the mixing of a system of large negatively charged particles with some small positively charged ones. As the particle number varies as the cube of the size, the small particles are likely to be present in larger numbers that the larger ones in the final mixture unless very small amounts are used. If the small particles are added to the large, the probable result will be a bridging aggregation with complete separation of the large particles from suspension. If the large are added to the small with good mixing, a system of large particles coated with small ones can be achieved. The order and mode of mixing in a hetero-coagulative system is just as important as it is with the mixing of adsorbing polymers to a dispersion.

4.3 Aggregate Structure

We have already seen how the hetero-coagulation of a clay leads to an open structure. This type of structure can space-fill, that is, occupy the whole volume available to the dispersion at low volume fractions. This has important implications for applications involving aggregated systems. For example, the rheological properties change dramatically so that handling can become difficult. Filtration may start as an easy separation process once a system is aggregated but the final 'de-watering' is limited and so subsequent drying can be a slow and expensive process. A relatively weak aggregation would be a preferable situation here so that collapse of the filter cake to a high-solids density could be achieved. In the case of ceramics, the rheology of the open structures is excellent for shape formation with minimal elastic recovery after yield at moderate to high stress. However, the open structures result in considerable shrinkage on drying and firing. These two examples illustrate clearly that the details of aggregate formation have great practical importance.

If the potential barrier to aggregation is reduced to zero, we have the situation where particle collisions are 'sticky'. This mode of aggregation is termed *diffusion-limited aggregation*. The simplest visualization of this is to consider the particles with a hard-sphere potential which has been modified to give a narrow but deep square attractive well close to the surface. Particles then collide and stick as they diffuse. The aggregates grow in size with an open-dendritic or fractal-type structure. Computer models generate this type of open branched type of structure and some careful experiments have confirmed these models. A fuller description may be found in Russel *et al.* [5] As these aggregates grow by accretion of 'sticky' particles, they grow into each other and span the available space. This point is known as the *percolation threshold*. At higher concentrations, denser structures result, and these are more difficult to define by a single parameter such as the 'fractal dimension'. In addition, although these structures are academically interesting, they are invariably modified in general applications. Systems are mixed during the addition of a coagulant and further 'shear-processed' during subsequent handling. The shear forces on these very large and fragile structures compact them to relatively high densities. For example, systems of monodisperse spherical particles can be compacted to random packing densities (i.e. $\varphi \sim 0.64$) in monodisperse spherical aggregates by shearing the coagulating system [8]. The point to remember here is that the strongly aggregated structures are *metastable* structures. The lowest energy configuration would be a very dense unit with the maximum number and/or area of contacts.

The structures that we observe then are a combination of processing and the strength of the attractive interaction. The attractive well can be controlled by the addition of material to the surface prior to coagulation. Non-ionic surfactants and polymers are excellent candidates for this, providing a steric barrier and limiting the aggregation to a weak flocculation. Excellent experimental studies of aggregate structures formed with weakly flocculated systems have been reported by Russel *et al.* [5] and Sperry [9]. A series of different sized polymer latices were mixed with different molecular weight water-soluble polymers to give depletion-flocculated systems. Phase-separation occurred at a value of $V_{min} \sim -2k_B T$. Stronger attractions resulted in rigid branched aggregates, while weaker attractions gave fluid-like aggregates more akin to a separated fluid phase. So, at some combination of attractive potential and thermal energy, equilibrium structures were attainable, but if $V_{min} \gg -k_B T$ non-equilibrium structures are formed. For phase-separation to occur, the time for the break-up of a pair of particles must be significantly longer than the time for more particles to collide with the doublet. However, if it is much longer than the time over which we are prepared to study the system, we will not wait long enough to see an equilibrium structure. The diffusion coefficient, D, of a particle is given by the Stokes–Einstein equation, as follows:

$$D = \frac{k_B T}{6\pi\eta_0 a} \tag{5.7}$$

where a is the particle radius and η_0 the viscosity of the medium. The diffusion constant has units of a flux, that is, $m^2\ s^{-1}$, and we can calculate the characteristic diffusional time as the time it takes for a particle to diffuse through a mean distance of a. Thus, the Einstein–Smoluchowski equation gives the time as:

$$t_D = \frac{a^2}{D} = \frac{6\pi\eta_0 a^3}{k_B T} \tag{5.8}$$

However, if there is an attractive potential the break-up rate will be slowed down as only a fraction of the particles will have sufficient thermal energy at any time to escape. The diffusion time must therefore be modified by the Boltzman factor to reflect this energy distribution:

$$t_D = \frac{6\pi\eta_0 a^3}{k_B T}\exp\left(-\frac{V_{min}}{k_B T}\right) \tag{5.9}$$

Figure 5.6 indicates how t_D increases with the attractive potential for particles of different size.

4.4 Slow Structural Changes

The usual laboratory timescale (our timescale) is 1 ms to 1 ks, but times which are very much longer than this cannot be neglected as products can be in

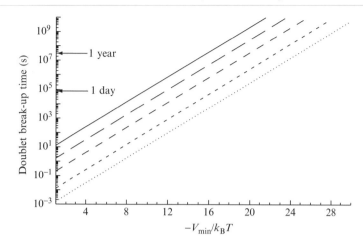

Figure 5.6. The time taken for doublets to separate as a function of particle radius and the depth of the attractive potential maximum: ($\cdots\cdots$) $a = 50$ nm; (- - - -) $a = 100$ nm; (– – – –) $a = 250$ nm; (——) $a = 500$ nm; (——) $a = 1000$ nm.

storage for several months. There are two points to note here. The first is that if new, 'sticky', particles arrive at a doublet before the doublet can break-up we can expect to observe phase-separation. We see how the break-up time increases rapidly with both particle size and attractive potential. The second point is more subtle as we can find changes occurring slowly with time. These are also related to the diffusive process. Consider a particular system with monodisperse spherical particles of 500 nm radius and weakly flocculated with an attractive potential of $V_{min} \sim -10k_BT$. At a volume fraction of ~ 0.3, we have a flocculated phase and experiments indicate that this is close to space-filling (see, for example, Goodwin and Hughes [10]). However, this is not the densest packing, which would be closer to face-centred cubic. From Figure 5.6, we see that the break-up time for a pair is ~ 2 h. What occurs is that particles in the lower-density zones diffuse to higher-density zones, i.e. they increase the number of nearest neighbours (coordination number) to move into a lower energy state. Migration in the opposite direction is clearly less favourable as each extra contact increases the total V_{min} by $10k_BT$. The final result is a slow change in local density followed by a collapse of the structure as it suddenly consolidates. This can take place over periods of time from hours to weeks, depending on the volume fraction (as this controls the local mean particle density or coordination number). We see an apparently stable system which, after a long incubation time, sediments rapidly with tracks being formed as the medium is forced up to the top of the bed.

Consolidation of suspension concentrates on storage is a common problem in systems of large dense particles.

5 THE RATE OF COAGULATION

5.1 Diffusion-Limited Aggregation

The initial rate of aggregation was first analysed quantitatively by von Smoluchowski who modelled the system as that of diffusing spherical particles which stick on collision contact but where the pair-potential is zero up to that contact. Detailed descriptions of the model for the diffusion behaviour can be found in Hunter [4] and Russel *et al.* [5]. A simplified approach is sufficient for our purposes here.

Figure 5.7 illustrates the geometry of this model. If r is the centre-to-centre distance between our reference sphere and an approaching particle, contact occurs when $r = 2a$ for a monodisperse system. The central particle is treated as a 'sink' so that no account is taken of any growth of the reference unit. This is satisfactory as the description is of the *initial* coagulation rate. As both particles are diffusing, the net diffusion coefficient is equal to $2D\,\mathrm{m}^2\,\mathrm{s}^{-1}$ [2]. The net velocity of an incoming particle is therefore $2D/a\,\mathrm{m}\,\mathrm{s}^{-1}$. The surface area of our 'collision sphere' is $4\pi(2a)^2$. The flux through the collision sphere, if there are initially N_p particles per unit volume, is:

$$J_\mathrm{B} = N_\mathrm{p}\frac{2D}{a}4\pi(2a)^2 \qquad (5.10)$$

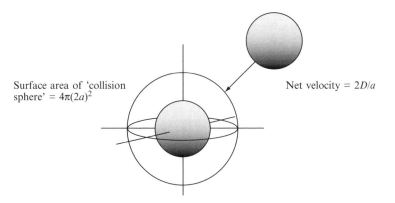

Surface area of 'collision sphere' = $4\pi(2a)^2$

Net velocity = $2D/a$

Figure 5.7. Illustration of Brownian collision of two identical particles of radius a with a diffusion constant of D.

As this process is occurring with each particle, the collision frequency due to Brownian diffusion is:

$$c_B = \frac{N_p J_B}{2} \qquad (5.11)$$

where the factor of '2' is introduced to prevent 'double-counting'. Substituting for D from Equation (5.7) into Equation (5.11) gives:

$$c_B = N_p^2 \frac{8 k_B T}{3 \eta_0} \qquad (5.12a)$$

As each collision results in coagulation, the initial coagulation rate that we should observe in a quiescent system is given by the following:

$$-\frac{dN_p}{dt} = N_p^2 \frac{8 k_B T}{3 \eta_0} \qquad (5.12b)$$

with the half-life of the rapid aggregation process being determined for this second-order rate equation from:

$$t_{\frac{1}{2}} = \frac{3 \eta_0}{4 k_B T N_p} \qquad (5.13)$$

However, the formation and aggregation of doublets and larger multiplets should be included [3] and results in Equation (5.13) providing a poor estimate of the situation.

5.2 The Effect of the Continuous Phase

The fast aggregation rate is inversely proportional to the viscosity of the suspending medium. This rate appears to be directly proportional to the temperature but increases more rapidly than this as the viscosity of a liquid decreases exponentially with increasing temperature. Clearly, a high viscosity reduces the rate of aggregation but there is an additional effect of the particle being in a liquid medium. The range of the hydrodynamic disturbance around a moving particle scales with the particle radius. This means that as two particles approach the hydrodynamic interactions start to become significant as $r < 3a$. As $r \rightarrow 2a$, the reduction in the rate of approach becomes large with the liquid between the two particles having to drain out of the intervening space before the particles can come into contact. The hydrodynamic interaction acts in a similar fashion to a repulsion in that it slows the approach and hence extends the time taken for the formation of doublets. This problem has been analysed by Derjaguin[11] and Spielman[12], with the latter calculating the reduction in the initial aggregation rate as a function of the viscosity of the medium.

5.3 Potential-Limited Aggregation

The von Smoluchowski rate given in Equation (5.12b) will overestimate the rate if there is a repulsion between the particles. Fuchs tackled this problem and his analysis is discussed in detail in Hunter [4] and Russel *et al.*[5]. The pair-potential slows the approach of two particles. At any distance, the fraction of particles with thermal energy in excess of the potential at that distance is given by the Boltzman factor, i.e. $\exp(-V_T/k_B T)$. The flux through successive spherical shells as the particles approach is slowed from the simple collision case and only a fraction of the particles that encounter one another can approach close enough to stick. The fraction of the encounters that stick is $1/W$, where W is known as the *stability ratio*. We can express W as the ratio of the two fluxes as follows [4]:

$$W = 2a \int_{2a}^{\infty} \exp(V_T/k_B T) \frac{dr}{r^2} \tag{5.14}$$

Reerink and Overbeek [13] pointed out that the maximum in the pair-potential was the dominant factor in restricting the approach of particles as in the slow-coagulation regime the electrolyte concentration is such that the diffuse layer is very compressed. They showed that a useful approximation to the integral equation (5.14) was:

$$W \approx \frac{1}{2\kappa a} \exp(V_{max}/k_B T) \tag{5.15}$$

Figure 5.8(a) shows the pair-potentials calculated for silver bromide particles with a ζ-potential of $-50\,\text{mV}$ and a particle radius of 50 nm. Equation (5.15) was employed to estimate the stability ratio for the same particles using the pair-potentials from Figure 5.8(a), with the results being shown in Figure 5.8(b). However, the height of the primary maximum relative to the secondary minimum was used in the calculation as this would be the energy required for the particles to move from a secondary minimum flocculation to a primary minimum coagulated state. It is clear from Figure 5.8(b) that the stability ratio changes very rapidly with electrolyte concentration. The result is that we will

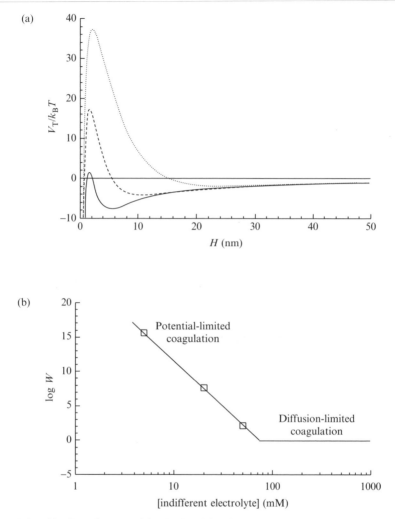

Figure 5.8. (a) The pair-potential calculated for silver bromide with a particle radius of 50 nm and a ζ-potential of -50 mV: ($\cdots\cdots$) $\kappa a = 11.6$, $V_{max} = 37.3k_BT$; (----) $\kappa a = 23.2$, $V_{max} = 17.4k_BT$; (——) $\kappa a = 36.7$, $V_{max} = 1.5k_BT$. (b) The stability ratio calculated for silver bromide as a function of the concentration of a 1:1 indifferent electrolyte (using Equation (5.15)).

usually just notice a change from a stable dispersion to a rapidly aggregated one under normal laboratory conditions. So, although it might appear that the electrolyte concentration of a sample is low enough for the stability to be adequate, the slow coagulation process is the type of problem that occurs during storage.

6 AGGREGATION IN FLOWING DISPERSIONS

There are two effects of a flow field on the aggregation behaviour of dispersions that we need to consider. The first is that the collision frequency is increased and this effect was first analysed by von Smoluchowski (see, for example, Hunter [4] and Russel *et al.* [5]). This just has an effect on the rate of aggregation but the hydrodynamic contribution to the interaction should also be considered [5]. This latter effect can cause a stable dispersion to aggregate under certain conditions. Aggregation under quiescent conditions is termed *orthokinetic* coagulation where collisions are solely due to Brownian motion. In a strong flow field, where the collisions are controlled by the shear field, the aggregation is termed *perikinetic* coagulation.

6.1 The Effect of Flow on Collision Frequency

This discussion will be restricted to laminar flow. In a high-speed mixer, the flow is turbulent with chaotic vortices so that the particles are subjected to a wide and unpredictable range of hydrodynamic forces and a much more complex treatment would be required.

In a mixed or flowing system, there is a variation of fluid velocity with position in the fluid. This is known as a *shear field*. With simple shear, the fluid is moving at a uniform velocity in, say the x–y plane, with a change in velocity in the z-direction. This is illustrated in Figure 5.9.

In the absence of Brownian motion, a particle will move at the velocity of the liquid at the plane coincident with the centre of the particle, v_p, which is a distance z_p from the reference x–y plane:

$$v_p = z_p \frac{\mathrm{d}v}{\mathrm{d}z} \tag{5.16}$$

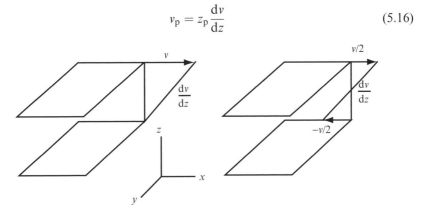

Figure 5.9. Illustrations of a simple shear field, where both figures are mathematically equivalent with the same velocity gradient. The right-hand representation is used in this case as it emphasizes the symmetry of the collision process.

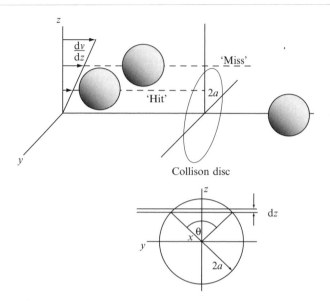

Figure 5.10. Illustration of collision trajectories for particles under simple shear flow. Note that there is symmetry on each side of the $x-y$ plane.

The geometry is illustrated in Figure 5.10. Any particle whose centre passes through the 'collision disc' with a radius of $2a$ collides with the reference particle on the x-axis. The problem is now straightforward as it reduces to calculating the fluid flux through the 'collision disc' and as we know that the particle concentration is N_p, we have the number of collisions with a reference particle. The fluid flux through the area element dz is given by.

$$f_z = \frac{dv}{dz} z dz 2y_z \tag{5.17a}$$

$$f_z = \frac{dv}{dz} [2a \cos(\theta/2)][-a \sin(\theta/2)][2a \sin(\theta/2)] \tag{5.17b}$$

Therefore, the total flux of particles through the disc is as follows:

$$f_p = N_p 2 \frac{dv}{dz} (2a^3) \int_\pi^0 [-\sin^2(\theta/2)\cos(\theta/2)] d\theta \tag{5.18}$$

As collisions occur with all of the particles, the total collision frequency due to flow, c_f, is the product of f_p and N_p, although we have to divide by two to prevent 'double-counting':

$$c_f = \frac{16}{3} N_p^2 a^3 \left(\frac{dv}{dz}\right) \tag{5.19}$$

However, as the particles approach in the shear field the hydrodynamic interactions cause the colliding pair to rotate and with the combination of the slowing of the approach due to liquid drainage [5,12], i.e. lubrication stresses, and Brownian motion, not all collisions will lead to aggregation. Equation (5.19) must be reduced by a factor α to account for this:

$$c_f = \alpha \frac{16}{3} N_p^2 a^3 \left(\frac{dv}{dz}\right) \tag{5.20}$$

This 'collision efficiency factor' is of order '1' and a typical value would be $\alpha \sim 0.8$. It is interesting to compare the Brownian collision frequency (c_B) from Equation (5.12a) with that due to flow in Equation (5.20):

$$\frac{c_f}{c_B} = \frac{2\alpha \eta_0 a^3}{k_B T} \frac{dv}{dz} \tag{5.21}$$

If the particles are dispersed in water at a temperature of 25 °C, the ratio in Equation (5.21) becomes:

$$\frac{c_f}{c_B} \approx 4 \times 10^{17} a^3 \frac{dv}{dz} \tag{5.22}$$

When we stir liquid in a beaker with a rod, the velocity gradient is in the range 1 to 10 s^{-1}: With a mechanical stirrer, 100 s^{-1} would be a reasonable value, while at the tip of a turbine in a large reactor one or two orders of magnitude higher could be possible. Hence, the particle radius a must be $< 1 \, \mu$m if even slow mixing can be disregarded.

This description of the collision process does not include the details of the collision trajectories which are governed by the hydrodynamic interactions. A more detailed description should also include the effects of interparticle repulsion and attraction, as well as the contact time [3].

6.2 The Effect of Flow on the Interaction Force

The above description of the collision frequency due to von Smoluchowski is useful if we are looking at particle coagulation rates in the absence of a potential barrier. When we consider the case of potential-limited aggregation, we have to consider the contribution due to the hydrodynamic forces acting on the colliding pair. An excellent analysis of this situation has been carried

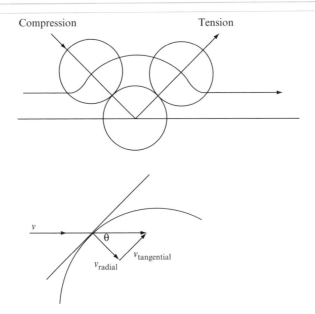

Figure 5.11. Illustration of the geometry of a colliding pair of particles with the maximum compression and tension at 45° to the shear plane.

out by Schowalter and is given in Russel *et al.* [5]. To illustrate the underlying concept, we will consider a simple model.

Figure 5.11 shows the forces acting on a collision doublet in simple shear. This figure shows the trajectory with the points at which the maximum compression and tension occur, that is, at values of $\theta = 45°$. The particles have the same radius a and the reference particle is at $z = 0$. The velocity of the streamline coincident with the centre of the colliding particle at the orientation giving the force maximum is:

$$v = \frac{\mathrm{d}v}{\mathrm{d}z} 2a \sin(45)$$

(5.23)

Now, the radial component of the Stoke's drag force on the particle is:

$$F_\mathrm{h} = 6\pi\eta_0 a v_\mathrm{radial} = 6\pi\eta_0 a \cos(45)$$

(5.24)

and so we may write:

$$F_\mathrm{h} = \pm\frac{\mathrm{d}v}{\mathrm{d}z} 6\pi\eta_0 2a^2 \sin(45)\cos(45)$$

(5.25a)

$$F_h = \pm \frac{dv}{dz} 6\pi\eta_0 a^2 \qquad (5.25b)$$

where the '$\pm$' indicates the compression or tension, respectively. It should be noted at this point that the trajectory illustrated in Figure 5.11 is a simplification. This trajectory would be altered by the colloidal forces on close approach. The details would depend on whether there was a net attraction or repulsion but the maxima and minima would still be at the positions shown and therefore Equation (5.25b) is satisfactory for our purposes.

It is possible to use Equation (5.25b) to indicate where the stability or instability boundaries are located for particular dispersions. To carry this out, all we have to do is to equate the interparticle forces at the maximum and minimum points on the force–distance curve. As an example, we shall consider a polystyrene latex system at a sodium chloride concentration of 50 mM. For the calculation, we shall choose a particle radius of 500 nm. The pair-potential can be calculated by using Equation (4.33a) for the repulsion and Equation (3.24b) for the attraction (see earlier). The interparticle force is given by:

$$F_T = -\frac{dV_T}{dH} \qquad (5.26a)$$

and so we have:

$$F_T = 2\pi\varepsilon_r\varepsilon_0\kappa a\psi_\delta \left(\frac{\exp(-\kappa H)}{1 + \exp(-\kappa H)}\right) - \frac{aA_{11}}{12H^2} \qquad (5.26b)$$

The value of κa is 368 and so we have a situation where the interparticle forces are changing in a region very close to the particle surface; hence, hydrodynamics control the trajectories until the particles are at very close distances. The force–distance curve for this system, with a ζ-potential of -40 mV, is shown in Figure 5.12, where $\zeta \cong \psi_\delta$ is assumed. The stability boundaries are calculated in terms of the value of the shear rate required to change the aggregation state at different values of the ζ-potential. The results obtained are shown in Figure 5.13. There are several features of note in this stability diagram. At ζ-potentials less than 20 mV, the dispersion is coagulated at all values of the shear that are plotted. With a small increase in the potential above this value, the dispersion is aggregated, but is flocculated and not coagulated at low shear rates. However, at shear rates of the order of 10^5 s^{-1} the hydrodynamic forces are sufficient to cause the dispersion to form doublets which are coagulated. It is also interesting to note that the shear forces on this particle size, ionic strength and diffuse layer potential combination will only break down the doublets flocculated in the secondary minimum when shear rates of the order of 10^3 s^{-1} are reached. Although shear

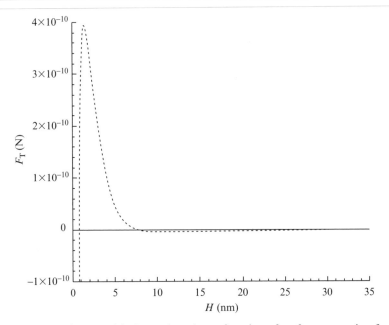

Figure 5.12. The interparticle force plotted as a function of surface separation for two polystyrene latex particles: $a = 500\,\text{nm}$; ζ-potential $= -40\,\text{mV}$; 1:1 electrolyte concentration $= 50\,\text{mM}$: $F_{\text{max}} = 3.96 \times 10^{-10}\,N$; $F_{\text{s, min}} = -3.80 \times 10^{-11}\,\text{N}$.

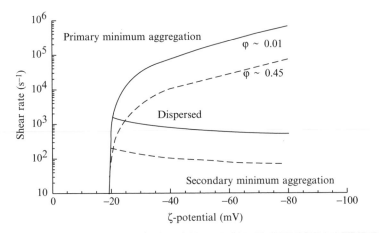

Figure 5.13. The stability map for polystyrene latexes as a function of shear rate and ζ-potential. The particle radius was taken as 500 nm, with the calculations being carried out for a 50 mM solution of a 1:1 electrolyte.

rates of this magnitude are readily attainable in a viscometer, they would represent a very high stirring rate with a paddle stirrer in a beaker on the laboratory bench, but are easily achieved in pumps and large reactors with turbine mixers. Equations (5.25a) and (5.25b) show that the shear forces increase as the square of the particle radius and so the stability boundaries drop rapidly with increasing particle size as the colloidal forces change more slowly with radius. Hence, particle with a radius of 3 or 4 μm are much more sensitive to shear-induced aggregation than particles of 0.3 or 0.4 μm in radius. It is also important to note that at electrolyte concentrations which are becoming high enough to be approaching a slow coagulation condition, the ζ-potential is unlikely to be as high as $-50\,mV$. This means that we are most interested in the steep part of the coagulation boundary where relatively low shear rates can be important. For example, localized high electrolyte concentrations can occur during the addition of a solution to a dispersion, even when the systems is stirred, and so attention must be paid to the mixing and addition rates.

In many situations, the volume fraction of the dispersion is maintained as high as it is possible in order to retain suitable handling properties, that is good heat and mass transfer properties. Hence, when the hydrodynamic forces on a pair of particles are considered, it is really the 'multi-body' hydrodynamic forces that are critical. As a first-order approach, we can take the 'effective medium' or mean-field approach and use the viscosity of the dispersion to calculate the local force. To illustrate this, we shall use the high-shear result for the viscosity of a dispersion of monodispersed hard spheres. The details of the derivation are described in the discussion on the rheology of dispersions later in this volume, and so here we will just give the result obtained, as follows:

$$\eta = \eta_0 \left(1 - \frac{\varphi}{0.605}\right)^{-1.513} \tag{5.27}$$

The stability boundaries shown in Figure 5.13 by the dashed lines were calculated by using a volume fraction of 0.45. Clearly, the boundaries drop in proportion to the viscous forces as expected and so it is easier to reduce the breakup of flocculated pairs. Thus, a larger fraction of the stable area is occupied by particles that we would term 'dispersed'. However, the important thing to observe is that the boundary is moving away from the $-20\,mV$ threshold more rapidly, therefore indicating more shear sensitivity. Indeed, we now see that close to the $-20\,mV$ threshold for rapid coagulation, flocculated particles will become coagulated at only moderate shear rates with there being no opportunity to separate them. At first sight, this may seem an unimportant point, but we frequently have concentrated dispersions to which we wish to add stabilizers, such as a surfactant, for example, and of course, mix with the addition of a stabilizer solution. Once coagulated, separating the particles to allow the stabilizers to adsorb and function correctly may be impossible.

At this point of the discussion, the question naturally arises concerning the break-up of coagulated particles, as Equations (5.25a) and (5.25b) give an estimation of the hydrodynamic forces available. This is, in part, the same problem as we see in the dispersion of dry powders into a liquid medium and the break-up of liquid drops in emulsification, although both of these processes have additional factors. In the case of the 're-dispersion' of coagulated particles, the problem is to estimate the depth of the primary minimum as the shear force has to be sufficient to overcome the attraction and pull the particles apart to a distance equivalent to the maximum in the curve. Here, the problem comes down to details of molecular dimensions, so that the hydration of bound ions and their dimensions, for example, must control the approach. We could take a cut-off distance of, say 0.5 nm, as a typical minimum separation and use this to calculate a re-stabilization boundary. The force–distance curve is very steep at this point and so the calculation would not be of great practical use. Another problem is that on this scale the details of the surface topography, distribution of charges, etc. become important. Finally, experimentation to either confirm the result or to provide a basis for estimation becomes difficult as the shear rates required would invariably involve turbulent flow fields which are chaotic and difficult to describe.

REFERENCES

1. B. V. Derjaguin and L. Landau, *Acta Physicochim.*, **14**, 633 (1941).
2. E. J. W. Verwey and J. Th. G. Overbeek, *Theory of the Stability of Lyophobic Colloids*, Elsevier, Amsterdam, 1948.
3. J. Th. G. Overbeek, *Pure Appl. Chem.*, **52**, 1151 (1980).
4. R. J. Hunter, *Foundations of Colloid Science*, Vol. I, Oxford University Press, Oxford, UK, 1987.
5. W. B. Russel, D. A. Saville and W. Schowalter, *Colloidal Dispersions*, Cambridge University Press, Cambridge, UK, 1989.
6. E. Matijévic, *J. Colloid Interface Sci.*, **43**, 217 (1973).
7. J. Gregory, 'Polymer Adsorption and Flocculation', in *Industrial Water Soluble Polymers*, C. A. Finch (Ed.), The Royal Society of Chemistry, Cambridge, UK, 1996, pp. 62–75.
8. J. W. Goodwin and J. Mercer-Chalmers, 'Flow Induced Aggregation of Colloidal Dispersions', in *Modern Aspects of Colloidal Dispersions*, R. H. Ottewill and A. R. Rennie (Eds), Kluwer Academic Publishers, Dordrecht, The Netherlands, 1998, pp. 61–75.
9. (a) P. R. Sperry, *J. Colloid Interface Sci.*, **87**, 375 (1982); (b) P. R. Sperry, *J. Colloid Interface Sci.*, **99**, 97 (1984).

10. J. W. Goodwin and R. W. Hughes, *Rheology for Chemists – An Introduction*, The Royal Society of Chemistry, Cambridge, UK, 2000.
11. B. V. Derjaguin, *Discuss. Faraday Soc.*, **42**, 317 (1966).
12. L. A. Spielman, *J. Colloid Interface Sci.*, **33**, 562 (1970).
13. H. Reerink and J. Th. G. Overbeek, *Discuss. Faraday Soc.*, **18**, 74 (1954).

Chapter 6

The Wetting of Surfaces by Liquids

1 INTRODUCTION

At first sight, the wetting of a macroscopic surface by a liquid appears to be unrelated to the normal colloid discussion. It is, however, intimately related to the detailed surface chemistry and physics which determine the behaviour of particles. The same intermolecular forces are involved and the adsorption of macromolecules or surfactants is usually handled as though it was occurring on a macroscopic surface.

There are a large number of situations in which the manner and rate that a surface is wetted by a liquid are of major importance. Many colloidal materials are obtained as dry powders which are then dispersed in a liquid medium. Pigments and fillers immediately come to mind. The extraction of oil from porous rock by replacing it through pumping water into the well is another case. We could pick many others from commercially important areas such as the enhancement of mineral ores via flotation to the more apparently mundane operations such as washing clothes or maintaining visibility through widows in wet conditions. Hence, this is an area of study which fits naturally into problems of the application of colloidal systems, as we shall see in the following.

In Chapter 1, the concept of surface tension was introduced in terms of the Helmholtz free energy per unit area of surface (Equation (1.15)) which can be re-stated as follows:

$$\gamma_s = \frac{F_s}{A_s} - \sum_i \mu_{is} n_{is} \tag{6.1}$$

Colloids and Interfaces with Surfactants and Polymers – An Introduction   J. W. Goodwin
© 2004 John Wiley & Sons, Ltd   ISBN: 0-470-84142-7 (HB)   ISBN: 0-470-84143-5 (PB)

where the subscript 's' denotes the surface, F is the free energy and A the area, with μ and n being, respectively, the chemical potential and number of moles of each species in equilibrium. This formal definition helps to focus our attention on the details of the intermolecular forces, the molecular composition of the interfacial region and where we draw the 'Gibb's dividing surface' as our description of the actual location of the surface. However, the experimental observations that we make, and the interpretations that we will discuss in this chapter, treat surfaces as macroscopic with well-defined locations and analyse the force balances by using classical mechanics. It is possible to achieve a great deal with this approach but we must always keep in mind adsorption processes and that the surfaces should be in equilibrium unless we are specifically looking at rate processes.

2 THE CONTACT ANGLE

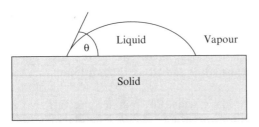

Figure 6.1. The contact angle, θ, formed by a drop of liquid on a solid surface.

When a small amount of liquid is placed on the surface of a solid it forms a drop which covers a limited area of the surface. This is illustrated in Figure 6.1 in which the contact angle, θ, is the angle between the tangent to the liquid at the contact line and the solid surface[1]. The limiting condition is that $0° < \theta < 180°$. If we take the condition where the contact angle approaches zero, we have a surface that is completely wetted by the liquid. Water on *clean* glass is an example of this. The glass is hydrophilic due to the silica surface, along with the large number of oxygen atoms and surface silanol groups which can hydrogen bond with the water surface. Mercury on a polytetrafluorethylene (PTFE) surface forms drops with a contact angle of about 150° and this can be considered to be complete non-wetting of the surface. If the value of the contact angle is $\geq$ 90°, the droplet does not spread readily. If the volume in the drop is changed, we can observe that the line at the three-phase junction, the *wetting line*, moves with some difficulty. If liquid is withdrawn, the angle is reduced from the value that we see when liquid is added. These experiments have to be carried out slowly and carefully so that we are as close to an equilibrium value as possible. The angle obtained as we just finish expanding the drop is known as the *advancing contact angle*, θ_A, while the one that is observed as the liquid has just been withdrawn is the *receding contact angle*, θ_R. These angle change if they are measured while the wetting line is in

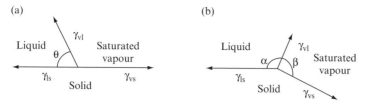

Figure 6.2. The balance of interfacial tensions at the vapour–liquid–solid contact line shown for (a) a flat surface, and (b) a 'non-flat' surface.

motion. They are then known as the *dynamic contact angles* and merit a discussion in some detail due to their importance in many coating and printing processes.

2.1 The Young Equation

In the illustration of the liquid drop shown in Figure 6.1, the contact line will move until equilibrium is established. We may describe this situation in terms of the force balance in the plane of the surface shown in Figure 6.2(a) and expressed as the vector addition:

$$\gamma_{vs} = \gamma_{ls} + \gamma_{vl} \cos \theta \tag{6.2}$$

Here γ_{12} is the *interfacial* tension between phases '1' and '2' (e.g. vapour (v) and solid (s)). (It is only referred to as the surface tension if it refers to the liquid/saturated vapour interface.) We should note that Equation (6.2) refers to a drop on a flat surface. If the surface is not flat, then the balance shown in Figure 6.2(b) gives [2, 3]:

$$\alpha + \beta \neq 180°: \quad \gamma_{vs} \geq \gamma_{ls} + \gamma_{lv} \cos \beta; \; \gamma_{ls} \geq \gamma_{vs} + \gamma_{lv} \cos \alpha \tag{6.3}$$

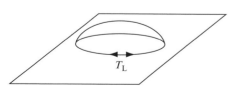

Figure 6.3. Schematic of the line tension, T_L, acting along the contact line of a liquid drop.

However, the contact line is not straight – it is curved and the radius of curvature is important as this results in what is known as the *line tension*, T_L, which acts in opposition to the expansion of the drop (as illustrated in Figure 6.3). Equation (6.2) should therefore be written for a drop of radius r as follows:

$$\gamma_{vs} = \gamma_{ls} + \gamma_{vl} \cos \theta + \frac{T_L}{r} \qquad (6.4)$$

The line tension is not very large, being typically of the order of $T_L \sim 10^{-11}$ N, with the value of $T_L/r > 1$ mN m^{-1} making a significant contribution. Therefore, we need only use Equation (6.4) for very small systems such as capillary condensation in porous solids.

3 METHODS FOR THE MEASUREMENT OF CONTACT ANGLE

There are several methods available for measuring the contact angle and suitable equipment may be purchased 'off the shelf'. The technique chosen will depend on the surface. For example, if it is a crystal face one of the first two methods described below would be suitable when using drops or bubbles. If it is a thin element which has had some form of surface treatment, one of the other methods may be easier. In all cases, the roughness of the surface is an issue which must be addressed.

3.1 Sessile Drop

In this approach, a drop of liquid is placed on the surface for which the contact angle is required. A syringe is then used – either to add liquid to give the value of the advancing angle, or to withdraw liquid to give the value of the receding angle (Figure 6.4). A low-power microscope coupled to a computerized image analysis system is the most convenient method for this evaluation. The simplest method is to draw a tangent to the image on the screen at the contact point of the drop and the surface. Rather better, however, is to digitize the image of the curved surface, fit the points to a polynomial and compute the tangent to the curve at the contact point. The drop and experimental surface should be enclosed in a cell so that the drop is surrounded by saturated

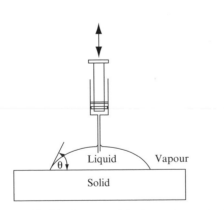

Figure 6.4. Schematic of the sessile drop experiment. In this approach, liquid is added or removed via the syringe and the drop profile is photographed.

vapour. This allows the surface of the solid to reach an adsorption equilibrium and to limit evaporation of the droplet. The drop must be sufficiently large that the syringe needle does not affect the curvature of the drop surface close to the contact line.

3.2 Captive Bubble

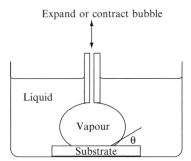

In this experiment, the solid is immersed in the liquid and a bubble is produced from a gas-tight syringe and brought into contact with the surface (Figure 6.5). Again, a low-power microscope coupled to a camera and a computerized imaging system is the most convenient method of determining the angle. In this case, expansion of the bubble enables the receding angle to be recorded, while contraction gives the advancing angle. This method has an advantage in that a saturated vapour is readily achieved.

Figure 6.5. Schematic of the captive bubble experiment. Expanding the bubble gives θ_R, while contraction provides θ_A.

3.3 Wilhelmy Plate

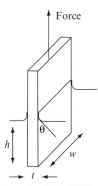

This is caried out by employing the same equipment that can be used to determine the surface tension of a liquid. In this mode, the plate must be fully wetted and a contact angle of $\sim 0°$ is assumed. However, if the surface tension of the liquid is known, the data can be used to calculate the contact angle. The experimental arrangement is shown in Figure 6.6. The change in force recorded by the

Figure 6.6. Schematic of the Wilhelmy plate experiment, where h is the depth of immersion of a plate of width w, with a thickness t. The net force is measured with a balance.

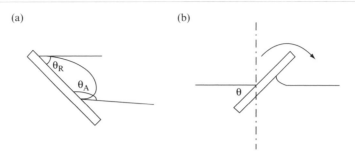

Figure 6.7. Schematics of the two tilted plate methods for measuring the contact angle.

balance on partial immersion of the plate is equal to the surface tension force minus the bouyancy of the immersed section of the plate, as follows:

$$\text{force} = 2(w + t)\gamma_{vl}\cos\,\theta - wth\rho_l \tag{6.5}$$

where w is the width of the plate of thickness t, and so $2(w + t)$ is the length of the contact line; h is the depth immersed while the liquid density is ρ_l. The advancing and receding angles are obtained by altering the depth of immersion.

3.4   Tilted Plate Methods

There are two methods which use tilted plates, as shown in Figure 6.7. When a drop is placed on an inclined plane, and the tilt of the plane is increased until the drop just starts to move, the advancing and receding angles may be obtained in one experiment (Figure 6.7(a)). This sounds easy but it requires a continuous recording of the image to determine the point to take the measurement and the technique is particularly prone to vibrations which lead to significant experimental scatter. The second technique, illustrated in Figure 6.7(b), also involves a tilted plate and makes use of the observation that the meniscus will appear flat when the plate makes an angle of $180 - \theta$ with the surface. Advancing and receding angles can only be obtained by changing the immersion depth of the plate.

4   CONTACT ANGLE HYSTERESIS

The value of the contact angle obtained experimentally is found to depend on whether the liquid is advancing over the surface or receding, and $\theta_A \geq \theta_R$. We may also observe ageing effects. The first thing to consider is the nature of the surface. Although we may form a fresh surface, the nature of that surface rapidly changes by the adsorption of material from the atmosphere. This may be physical adsorption of water or hydrocarbons, or chemical adsorption to give an oxide layer on metals or water, as in the formation of surface silanol groups on freshly prepared silica. Recall from Chapter 3 that the surface tension tells us about the surface energy and for 'atomic liquids' this can be calculated from the London dispersion forces, while the hydrogen bonding in liquids such as water makes the dominant contribution. With solids, things can be more complex as there can be residual strains present, but solids have much higher surface energies (and tensions) than liquids. Nevertheless, the adsorption on the surface can make a very large difference. Isrealachvili [4] quotes data for cleaved mica which illustrate this well. For mica, freshly cleaved under high vacuum, the value is $4.5 \, \mathrm{J \, m^{-2}}$ and this falls to $0.3 \, \mathrm{J \, m^{-2}}$ when it is cleaved under normal laboratory conditions and water can adsorb. A thin liquid layer can form on the solid surface and we might think that the contact angle would become very small, as the liquid would in effect be in contact with itself. The wetting film has to be thick for this to be the case. With thin films such as monolayers, the underlying intermolecular forces are still evident. Of course, when we make measurements of advancing angles, there may only be a monolayer or so of liquid adsorbed on the solid, but when we reverse the process and withdraw liquid and rapidly measure the receding angle, a much thicker film may be present which only thins slowly.

However, there are other reasons for the observed hysteresis. Surface heterogeneity can result from polycrystallinity or variation in local composition. When the wetting line moves over a heterogeneous surface, it will stick as it passes a boundary from a lower-energy to a higher-energy region. This will hold back the contact line, thus increasing the advancing angle and reducing the receding angle. Surface roughness is an important cause of hysteresis. The schematic shown in Figure 6.8 illustrates the effect. When the angle is measured, the surface is taken as a plane through the undulations, although the liquid must form the angle to the local surface, which may be at an angle to the viewing plane. As shown in this figure, if the surface is at an angle α to the measurement or viewing plane, the receding angle will be apparently

Figure 6.8. Schematic of a drop on a rough surface. Here, we measure an 'apparent' contact angle, and hence the differences between the advancing and receding angles, even though the true contact angle does not alter ($\sim 90°$ in the figure).

reduced by α and the advancing angle increased by α at the points shown. What will happen is that the drop will move in 'jumps' from one metastable configuration to the next. Wenzel[5] has defined a 'roughness factor', R_f, in terms of the ratio of the actual surface area to the apparent geometric area for which the apparent contact angle was θ', and so Young's equation now becomes:

$$R_f(\gamma_{vs} - \gamma_{ls}) = \gamma_{vl} \cos \theta' \qquad (6.6a)$$

and so:

$$R_f = \frac{\cos \theta'}{\cos \theta} \qquad (6.6b)$$

The dimension scale of the surface heterogeneity or roughness is important. A dimension of at least 100 nm appears to be required. This means, for example, that optically smooth surfaces will not show much hysteresis due to roughness.

5   SPREADING

The surface tension of a pure liquid was defined as the surface energy of that liquid. That is, it is the energy required to generate the unit area of a new surface. The *work of cohesion*, W_{ll}, of a liquid is the work that would be required to separate two 'slabs' of the liquid, as shown in Figure 6.9(a):

$$W_{ll} = 2\gamma_{vl} \qquad (6.7)$$

The *work of adhesion*, W_{ls}, is the work required to separate two different materials (shown in Figure 6.9(b)). This is calculated as the sum of the energies of the new surfaces that are formed, minus that of the interface that is lost, as follows:

$$W_{ls} = \gamma_{vs} + \gamma_{vl} - \gamma_{ls} \qquad (6.8)$$

Figure 6.9(c) shows the spreading in similar terms. We may define the *equilibrium spreading coefficient*, S, in terms of the work of adhesion and work of cohesion. We have to do work to create the new liquid surfaces and then this is offset by the work of adhesion of the liquid to the vapour/solid surface:

$$S_{vls} = W_{ls} - W_{ll} \qquad (6.9a)$$

$$S_{vls} = \gamma_{vs} - \gamma_{vl} - \gamma_{ls} \qquad (6.9b)$$

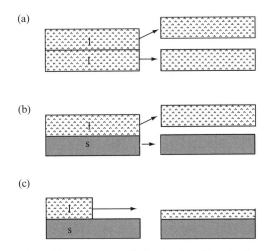

Figure 6.9. Schematics of the processes of (a) cohesion, (b) adhesion, and (c) spreading.

We may then use the Young equation (Equation (6.2)) to put both the work of adhesion and the spreading coefficient in terms of readily measured quantities such as the contact angle:

$$W_{ls} = \gamma_{vl}(\cos\,\theta + 1) \tag{6.10}$$

This is known as the Dupré equation, and the spreading coefficient is given by the following:

$$S_{vls} = \gamma_{vl}(\cos\,\theta - 1) \tag{6.11}$$

and, of course, the contact angle can be written in terms of the ratio of the work of adhesion and that of cohesion, as follows:

$$\cos\,\theta = 2\frac{W_{ls}}{W_{vl}} - 1 \tag{6.12}$$

It is instructive to note the following limits:

$$
\begin{aligned}
W_{ls} = W_{vl}; &\quad \cos\,\theta = 1, \text{ and so } \theta = 0° \\
W_{ls} = \frac{W_{vl}}{2}; &\quad \cos\,\theta = 0, \text{ and so } \theta = 90° \\
W_{ls} \ll W_{vl}; &\quad \cos\,\theta \rightarrow -1, \text{ and so } \theta \rightarrow 180°
\end{aligned} \tag{6.13}
$$

Of course, this is no surprise as the work of cohesion and adhesion are calculated from the surface energies, which are a reflection of the intermolecular forces. For non-polar liquids, the London dispersion forces provide a good estimate of the surface energy via the appropriate Hamaker constants. Girifalco and Good[6] and Fowkes[7] have observed that as the geometric mean gave a reasonable approximation for the combined Hamaker constant, the geometric mean of the work of cohesion, and hence the surface tensions, could be used to calculate the interfacial tension. So, for two liquids with surface tensions equal to γ_1 and γ_2, the interfacial tension, γ_{12}, from Equation (6.8), is as follows:

$$\gamma_{12} = \frac{W_{11}}{2} + \frac{W_{22}}{2} - W_{12} \tag{6.14}$$

or:

$$\gamma_{12} = \gamma_1 + \gamma_2 - W_{12} \tag{6.15}$$

Now:

$$W_{12} \approx \sqrt{W_{11} W_{22}} = 2\sqrt{\gamma_1^d \gamma_2^d} \tag{6.16}$$

and:

$$\gamma_{12} \approx \gamma_1 + \gamma_2 - 2\sqrt{\gamma_1^d \gamma_2^d} \tag{6.17}$$

where the superscript 'd' is used to indicate the dispersion force contribution to the surface tension. For example, the surface tension of water has a dispersion force contribution and a larger H-bonding contribution, thus $\gamma_{H_2O} = \gamma_{H_2O}^d + \gamma_{H_2O}^H = 22 + 50.5 = 72.5\,\text{mN m}^{-1}$. So, by using Equation (6.17) for octadecane on water we can calculate the interfacial tension at 20°C as $\sim (28 + 73 - 2(28 \times 22)^{0.5})$, i.e. $\sim 51\,\text{mN m}^{-1}$, as the interaction across the interface is due to the dispersion forces.

When we are considering liquids wetting a solid and we combine Equations (6.10) and (6.16), we obtain the following:

$$\gamma_{vl}(\cos\,\theta + 1) = 2\sqrt{\gamma_{vl}^d \gamma_s^d} \tag{6.18}$$

Note that the surface tension of the solid is not modified by the vapour from the liquid in this expression, and so we must be careful in the application of Equation (6.18). Rearrangement of the latter gives the following:

$$\gamma_s^d = \frac{\gamma_{vl}^d(\cos\ \theta - 1)^2}{4} \qquad (6.19)$$

So, we may use measurements of the contact angle and the dispersion contribution to the surface tension of the liquid/vapour interface to estimate the solid surface tension. Good results have been obtained for non-polar materials. In many cases, the vapour from the liquid adsorbs on the surface of the solid and the surface pressure of this wetting film, Π_{slv}, needs to be considered. Therefore, we should write:

$$\gamma_s = \gamma_{ls} + \Pi_{vls} \qquad (6.20)$$

The surface pressure is most significant when the contact angle approaches zero and the surface is completely wetted by the liquid which spreads over the surface giving stable films at all thicknesses. When the contact angle is greater than zero, the thinning film produced as the liquid spreads becomes unstable and breaks up into droplets.

The *critical surface tension of wetting*, γ_C, can be defined by making use of Equation (6.19) as $\theta \to 0$; then $\gamma_s^d \to \gamma_{vl}^d = \gamma_C$. A plot of $\cos\ \theta$ versus $(\gamma_{vl}^d)^{-0.5}$ should be linear and the value of the surface tension found by extrapolating the line to $\cos\ \theta = 1$ gives an estimate of the surface tension of the solid. This is known as a Zisman plot[8] and has been used to determine the surface energies of solids with low surface energies. However, this is only an approximation as the surface pressure would have to be zero and that would not be the case as θ approaches zero.

5.1 The Adsorption of Surfactants

In many practical situations, surfactants are present as stabilizers, dispersants or wetting aids. Often, we will be concerned with the wetting of solids using aqueous solutions but that will not always be the case. The adsorption of the surfactant at each interface must be considered. Young's equation is a useful starting point and if it is written as:

$$\cos\ \theta = \frac{\gamma_{vs} - \gamma_{ls}}{\gamma_{vl}} \qquad (6.21)$$

it is immediately apparent that if either of the surface tension values at the liquid–solid or liquid–vapour interfaces decrease, then $\cos\theta$ will increase and so wetting is improved as this means that the contact angle moves towards zero. As the surfactant is in the liquid phase, adsorption at the solid–vapour interface may not be complete. For example, this would be the case if we were coating by using a blade or a roller technique. In other situations, this may not be the case as the solid may be in the liquid phase prior to the gas phase being introduced. The separation of minerals by *froth flotation* is an example here. In this process, the crushed mineral and its matrix are mixed with water and a surfactant solution is added (the 'collector'). Preferential adsorption onto the mineral fragments occurs and air is pumped into the solution to produce a froth. Additional surface active materials ('frothers') are used to reduce the liquid–vapour tension to enhance the frothing response. These have to be carefully chosen so that they do not change the contact angle too much. The contact angle at the mineral–liquid–vapour line is increased ($\cos \theta \rightarrow 0$) so that the desired mineral particles become attached to bubbles and are removed in the froth. There is a complex situation here as there is more than one type of solid surface present, as well as more than one surfactant being added with quite specific interactions with the surfaces occurring. As a simple example, let us consider the effect of a cationic surfactant such as hexadecyltrimethylammonium bromide ($C_{16}TAB$), at the glass (fused silica)–water–air interface. The silica surface is negatively charged at pH values of 2 or higher and the cationic head group produces strong adsorption of the surfactant at low solution concentrations. A monolayer of surfactant at the liquid–solid interface is formed by the time a solution concentration of $\sim 5 \times 10^{-5}$ M is reached. The contact angle of water on clean glass is $< 10°$ while with a monolayer of $C_{16}TAB$ this is $\sim 90°$. The cmc of the surfactant is $\sim 10^{-3}$ M, and at that concentration a bilayer of surfactant is adsorbed and the contact angle is again reduced to $< 10°$. At a concentration of 5×10^{-5} M, the adsorption of the surfactant at the air–liquid interface is low and has only reduced the surface tension by about 10%. The large change is to the solid surface which now consists of a surface of densely packed alkane chains causing both the values of γ_{vl} and γ_{ls} to drop dramatically, although with the former (to $\sim 27\,mN\,m^{-1}$) falling more than the latter. Wetting is again favoured by the ionogenic surface formed by the bilayer in combination with the reduced tension of the air–water surface as the cmc is approached.

6   CURVED SURFACES

The surface tension acts in the plane of the surface, and so if we have a small volume of liquid a spherical drop is formed. There is a pressure drop across the interface and we may calculate this by balancing the work done in attempting to reduce the surface area with that generated by the internal pressure. The change in the surface energy of a sphere of radius r is given by the following:

$$\gamma_{vl}dA(r) = \gamma_{vl}8\pi r dr \tag{6.22}$$

At equilibrium, this is balanced by the change in the pressure drop, thus:

$$4\pi r^2 \Delta P dr = 8\pi \gamma_{vl} r dr \qquad (6.23)$$

The result is the Young–Laplace equation [1]:

$$\Delta P = \frac{2\gamma_{vl}}{r} \qquad (6.24)$$

where r is the radius of the spherical drop. Equation (6.24) applies to any curved interface and it may be generalized for two different radii of curvature measured orthogonally, as follows:

$$\Delta P = \gamma_{vl}\left(\frac{1}{r_1} + \frac{1}{r_2}\right) \qquad (6.25)$$

(Note the convention is that the radius of curvature is measured in the liquid phase and so the sign convention is positive for a drop and negative for a bubble.) This means that the vapour pressure of a liquid in a small drop is higher than for a flat surface but is lower in a bubble than the flat surface. A consequence of this is that large drops grow at the expense of small ones as the material evaporates from the smaller and condenses on the larger drops.

7 CAPILLARITY

The behaviour of liquids in capillaries and pores is an important aspect of the wetting of surfaces. The wetting of rock in oil wells is one such example. A single capillary is a suitable starting point for the discussion. Figure 6.10(a) shows a capillary with a circular cross-section of radius r in which the liquid has risen to a height h. The liquid rises until the surface force which causes the liquid to spread on the surface is balanced by the gravitational force on the column of liquid. The pressure drop across the interface (the *capillary pressure*) is given by the Young–Laplace equation for a circular capillary:

$$\Delta P = hg(\rho_1 - \rho_v) = \frac{2\gamma_{vl}}{r} \qquad (6.26)$$

In nearly all cases, $\rho_1 \gg \rho_v$. When the contact angle is greater than zero, the radius of curvature of the meniscus, r_c in Figure 6.10(a), is $r/\cos\theta$ and Equation (6.26) then becomes:

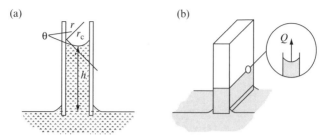

Figure 6.10. (a) Schematic of the capillary rise in a tube where the radius is r and the radius of curvature of the meniscus is r_c. (b) Schematic of the capillary rise in a porous block, where the volumetric flow rate in a 'pore' is Q.

$$h\rho_1 g = \frac{2\gamma_{vl}\cos\theta}{r} \tag{6.27}$$

This is an equilibrium measurement and provides another route to the determination of the contact angle. It should be noted that we have an inverse dependence of h on the radius of curvature. So, if we have a shape in which the radius of curvature changes we should expect to see a greater rise with the sharpest curvature. It is instructive to look at the behaviour of water in a small square glass optical cell. The meniscus is highest in the inside corners where there is a small radius of curvature. If the same cell is then stood in a dish of water, we can see that the wetting line on the outside is lowest at the outer corners (negative radius of curvature).

The imbibition of liquid into a capillary (or pore) is illustrated in Figure 6.10(b). The rate is often important in many situations. We start by using the Poiseuille equation for the flow of liquid in a tube:

$$Q = \frac{\Delta P \pi r^4}{8L\eta} \tag{6.28}$$

where L is the wetted length of a capillary of radius r, η is the viscosity of the liquid and Q is the volume flow rate, given by the following:

$$Q = \pi r^2 \frac{dL}{dt} \tag{6.29}$$

We may now substitute Equations (6.26) and (6.29) into (6.28) to give the Washburn equation, as follows:

$$\frac{dL}{dt} = \frac{r\gamma_{vl}\cos\theta}{4l\eta} \tag{6.30}$$

Integration of the above gives the distance the liquid penetrates in a given time:

$$\int_0^{L(t)} L dL = \frac{r\gamma_{vl} \cos \theta}{4\eta} \int_0^t dt \tag{6.31a}$$

$$L(t) = \left(\frac{r\gamma_{vl} \cos \theta}{4\eta} t\right)^{0.5} \tag{6.31b}$$

This result is for a single pore but it is readily adapted for a porous solid. If we have a porous block of material, for example, rock, the pores will vary in diameter and path. The effect can be included in one factor, the *tortuosity factor*, K:

$$L(t) = \left(\frac{r\gamma_{vl} \cos \theta}{4\eta K^2} t\right)^{0.5} \tag{6.32}$$

For a given porous system, r/K^2 is a constant. If the penetration is measured by using two or three different liquids which each have a value of $\cos \theta \sim 1$, but with different viscosities, the value of r/K^2 can be determined by using Equation (6.32).

The penetration of oils, water or surfactant solutions is of great importance in both the drilling for oil and gas, as well as in the recovery from mature wells. Some of the clay-based formations are soft, porous and easily wetted by water so that controlling the size of the well-bore can be difficult. The 'break-through' of water into the recovery well from well bores used to pump in replacement water is also a major problem. Polymers and surfactants can play an important role in controlling the contact angles and viscosities in these porous systems.

7.1 Dispersion of Powders

Another common situation concerning the penetration of liquid into a porous matrix is in the dispersion of dry powders in liquids. Many colloidal systems are dried and subsequently redispersed. The dry powders are easy to store and transport, and are also easy to handle in batch applications. Pigments, fillers and some foodstuffs are commonly handled in this way. When dried and ground, it is useful if the aggregates are macroscopic so that dust problems are minimized but large aggregates have to be wetted and broken up. The particles are held together by dispersion forces and in the dry state these are very strong as there is no intervening medium to mediate them and also the particles are in intimate contact in a deep attractive well. The first requirement is that the liquid must wet the particles. If the contact angle is low, the medium will be imbibed and the Laplace pressure is given by Equation (6.25) (suitably modified by the tortuosity factor K). The rate of penetration is given by the Washburn equation but note that as the aggregate is immersed in the fluid, the air inside is trapped and will give a back pressure and slow the penetration. As the pressure is higher than atmospheric, the gas will dissolve and allow the penetration to continue. However, this will be slower as the gas has to diffuse out of the aggregate through the imbibed liquid. This is clearly a slower process for large aggregates than small ones. As an example, we can take a pore radius of $0.1\,\mu m$, a surface tension for the wetting liquid of $40\,mN\,m^{-1}$ and a contact angle of $60°$. If $K \sim 1$, the Laplace pressure is $4\,atm$. Even if the pore penetration continues until the internal pressure reaches this value, it will be insufficient to break-up the aggregates, but will slow the wetting of the aggregates. Of course, we normally use mechanical action, that, is shear forces, to aid the break-up, by milling, for example. If the deformation rates (shear, extension, or both) are high and the concentration of aggregates is also high, there may be enough force to overcome the interparticle attraction. The hydrodynamic stress is then given by the following:

$$\frac{F_h}{a_f^2} \propto \eta(\varphi)\dot{\gamma} \qquad (6.33)$$

where F_h is the hydrodynamic force produced on an aggregate with a radius of a_f by a shear rate of $\dot{\gamma}$ (in s^{-1}), and $\eta(\varphi)$ is the viscosity of the concentrated suspension of aggregates. Deformation rates of 10^5 to $10^6\,s^{-1}$ are high, while much higher values are difficult to achieve. The value of the dispersion viscosity is of great importance here as stresses equivalent to several atmospheres would be required to overcome the dispersion forces. This is why it is much easier to disperse powders as a paste and subsequently dilute, rather than add all of the dispersion medium initially.

The wetting of the aggregate though is still an essential part as solvation of the surface allows repulsion to occur as well as reducing the attraction. Repulsion may be the development of electrical charge while it may also be the solvation of polymeric or protein layers. Both of these result in repulsion aiding break-up of the aggregates, as well as stabilizing the dispersed units to prevent re-agglomeration. This introduces another feature of capillary behaviour in porous systems. This is the tendency for vapours to condense in small pores and this will aid the wetting process.

7.2   The Kelvin Equation

The Young–Laplace equation indicates the pressure difference across a curved interface. The Kelvin equation enables us to calculate the change in vapour pressure due to the curvature of the interface. A clear derivation is given in Hunter [9], and we will give the outline here. For a pure liquid, at the curved interface at equilibrium the chemical potentials in the liquid and vapour phases are equal and the pressure drop is given by the Young–Laplace equation, as follows:

$$\mu_l = \mu_v$$

$$P_l - P_v = \frac{2\gamma_{vl}}{r} \tag{6.34}$$

We may make a small displacement from equilibrium and then we have:

$$d\mu_l = d\mu_v$$

$$dP_l - dP_v = d\left(\frac{2\gamma_{vl}}{r}\right) \tag{6.35}$$

where the changes in the chemical potentials are:

$$d\mu_v = \overline{V}_v dP_v - \overline{S}_v dT \tag{6.36a}$$

$$d\mu_l = \overline{V}_l dP_l - \overline{S}_l dT \tag{6.36b}$$

Here, the bars indicate the molar volumes and entropies. At constant temperature, and as we are close to equilibrium, we may write:

$$P_l = \frac{\overline{V}_v}{\overline{V}_l} P_v \tag{6.37}$$

Substituting this into Equation (6.34) yields:

$$\left(\frac{\overline{V}_v}{\overline{V}_l} - 1\right) dP_v = d\left(\frac{2\gamma_{vl}}{r}\right) \tag{6.38}$$

The boundary conditions that we need for the integration of Equation (6.38) are: $r \to \infty$; $P_v \to P^0$ (the vapour pressure of the liquid with a flat surface). Then, by taking note that as:

$$\frac{\overline{V}_v}{\overline{V}_l} \gg 1 \text{ and } \overline{V}_v \approx \frac{RT}{P_v}, \text{ then } \int_{P^0}^{P_v} \left(\frac{RT}{\overline{V}_l} \frac{dP_v}{P_v}\right) \approx \int_{\infty}^{r} d\left(\frac{2\gamma_{vl}}{r}\right) \tag{6.39}$$

the result for the vapour–liquid surface is given by:

$$\ln\left(\frac{P_v}{P^0}\right) \approx \frac{\overline{V}_1}{RT}\frac{2\gamma_{vl}}{r} \tag{6.40}$$

If the cross-section is oval instead of circular, then Equation (6.40) is written in terms of the major and minor radii, r_1 and r_2:

$$\ln\left(\frac{P_v}{P^0}\right) \approx \frac{\gamma_{vl}\overline{V}_1}{RT}\left(\frac{1}{r_1} + \frac{1}{r_2}\right) \tag{6.41}$$

If we consider a liquid drop, the radius of curvature (in the liquid) is positive and Equation (6.40) indicates that as r decreases, $P_v/P^0 > 1$, that is, the vapour pressure of the liquid in the drop becomes increasingly larger than that of the flat surface. This is in contrast to the case of a bubble where the radius of curvature (in the liquid) is negative and then decreasing the radius means that the vapour pressure becomes lower although the actual pressure (the capillary pressure) is higher (as given by the the Young–Laplace equation). The vapour pressure above the liquid in a capillary or a pore is also lower and this gives rise to the phenomenon of *capillary condensation*. In this case, vapours start to condense in the finest pores or crevices where there is a highly curved region prior to the saturation vapour pressure being reached.

This effect of condensation in regions with a sharp curvature can also be a problem with dry powders stored in a humid atmosphere. At the contact points between the particles (usually agglomerates), the curvature is such that capillary condensation occurs in the contact zones. The force exerted by the liquid can be large enough to make the powder particles stick quite firmly together and we recognize this as 'caking'. It is a useful exercise to calculate the force, assuming that the particles are smooth spheres completely wetted by the liquid phase [9]. The geometry is illustrated in Figure 6.11, where the radius of the particles is R and the radii of curvature of the liquid wetting the particles are r_1 and r_2. The Laplace pressure is, as given earlier:

$$\Delta P = \gamma_{vl}\left(\frac{1}{r_1} - \frac{1}{r_2}\right) \tag{6.25}$$

This gives the force F_{cc}, holding the surfaces together, as follows:

$$F_{cc} = \Delta P \pi r_2^2 = \pi\gamma_{vl}\left(\frac{r_2^2}{r_1} - r_2\right) \tag{6.42}$$

Now as $r_2 > r_1$, and noting from Figure 6.11 that $2R \approx r_2^2/r_1$:

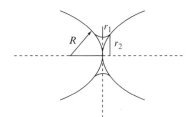

$$F_{cc} \approx 2\pi R \gamma_{vl} \qquad (6.43)$$

7.3 Solid Particles Solubility

It is interesting to note that the Kelvin equation can also be used to describe the change in solubility of a solid particle in a liquid as the particle size is reduced. The solubility increases as the size is reduced in the same manner as the vapour pressure of the liquid in a drop increases as

Figure 6.11. Illustration of liquid condensation in the contact region between particles in a powder to give a liquid bridge, where R is the radius of curvature of the particle surface, and r_1 and r_2 are the radii of curvature of the meniscus (Note that r_2 is negative in this figure.

the size is reduced. This means that larger particles grow by the transference of material from smaller particles. This is known as *Ostwald ripening* and the analogous form of the Kelvin equation is:

$$\ln \left(\frac{s(r)}{s(r \rightarrow \infty)} \right) \approx \frac{2\gamma_{ls} \overline{V}_s}{rRT} \qquad (6.44)$$

where $s(r)$ is the solubility of a solid with a radius r. However, the equivalent of capillary condensation also occurs where two crystals make contact (in a sedimented bed, for example) and then they fuse together and make redispersion impossible.

8 TEMPERATURE EFFECTS

As a material is heated we observe melting and then boiling of the liquid phase to produce the vapour. In some cases, we just observe sublimation of the solid. The surface tension of a material is due to the intermolecular forces in the material as we have already discussed. The surface tension then reflects the decreasing attraction and decreases with increasing temperature. This effect is larger in the liquid phase when compared to the corresponding solid and occurs for both simple molecular liquids where the intermolecular forces are dispersion forces or for those with other contributions. For example, the decrease in hydrogen bonding in water as the temperature is increased towards $100\,^{\circ}C$ is very familiar to us.

When we consider the equilibrium expressed by the Young equation (Equation (6.2)), we note that the change in the tension at the liquid–vapour interface will be greater than the difference in the changes of the tensions at the solid–vapour and solid–liquid interfaces. For equilibrium to be maintained, the contact angle must decrease, that is, $\cos\theta$ increases. This means that the surface we are considering is more easily wetted at higher temperatures. Adamson[1] provides a more extensive discussion on the effect of temperature on contact angle. We shall make use of this change to provide an estimate of the enthalpy of the wetting process.

8.1 The Heat of Wetting

The free energy change on completely immersing a solid in a liquid can be estimated from the difference in tensions as we replace the vapour–solid interface by a liquid–solid interface:

$$-\Delta G_w = \gamma_{vs} - \gamma_{ls} \tag{6.45}$$

We must keep in mind that Equation (6.45) is implying that the solid is uniform throughout and we can consider it to be in an equilibrium state. Thus, we are assuming that the tensions are a good approximation to the surface free energy of the solid. We associate the temperature-dependence with the change in entropy, and so we may take the approach of Harkins and Jura[10] in their work using heats of immersion:

$$\Delta H_w = \Delta G_w + T\Delta S_w \tag{6.46}$$

and by using Equation (6.45) with the variation of each tension with temperature, we have:

$$\Delta H_w = \left(\gamma_{ls} - T\frac{d\gamma_{ls}}{dT}\right) - \left(\gamma_{vs} - T\frac{d\gamma_{vs}}{dT}\right) \tag{6.47}$$

Grouping the like terms:

$$\Delta H_w = -\left[(\gamma_{vs} - \gamma_{ls}) - T\left(\frac{d\gamma_{vs}}{dT} - \frac{d\gamma_{ls}}{dT}\right)\right] \tag{6.48}$$

We may use the Young equation to put this in terms of the contact angle and the vapour–liquid tension:

$$\Delta H_w = -\left[\gamma_{vl}\cos\theta - T\frac{d(\gamma_{vl}\cos\theta)}{dT}\right] \tag{6.49}$$

This enables us to estimate the heat of wetting simply by measuring the changes in contact angle and the surface tension of the liquid as a function of temperature. In many cases, this is a simpler experiment than using calorimetry as the heat change is so small. However, the precision of the contact angle measurements is only a few degrees, and we must recognize the approximation in equating the surface tension of the solid with the surface free energy.

8.2 Dynamic Contact Angles

Thus far, the discussion in this chapter has been restricted to static processes which, if not in equilibrium, are in a *metastable* state. During the coating of surfaces, often at very high rates, the dynamics are important. The first issue to consider is the tensions at the liquid interfaces. If surfactants have been included in the formulation of a coating, the kinetics of the adsorption processes at both the liquid–solid and the liquid–vapour interfaces are important. As the timescale for diffusion to the interface is approached by the timescale of the creation of the new interfaces, adsorption will not be completed. As the coating rates are increased, the tensions approach those due to the major liquid component. In practice, this is not often a factor though as the surfactant concentrations are frequently around the critical micelle concentrations so as to ensure sufficient material to provide the colloidal stability of the particulates in the system.

   A more universal problem is that of the hydrodynamic or viscous stress. The viscous force acts in the same direction as the solid–liquid surface tension, that is, in the same direction in which the surface is moving or in the opposite direction to the motion of the coating implement whether a blade, brush or roller. Figure 6.12 provides a simplified schematic for the coating of a moving tape by using a blade to determine the thickness of the coat. At rest, we can use the Young equation (Equation (6.2)) for the force balance by using the advancing contact angle, as follows:

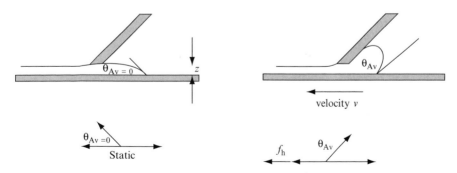

Figure 6.12. Schematic representation of a coating, using a blade to determine the film thickness.

$$\gamma_{vs} = \gamma_{ls} + \gamma_{vl} \cos \theta_{Av \to 0} \tag{6.2}$$

and when the coating is being applied we must include the viscous force term in the new force balance:

$$\gamma_{vs} = \gamma_{ls} + \gamma_{vl} \cos \theta_{Av} + f_h \tag{6.50}$$

where f_h is the hydrodynamic force per unit length of interface moving at a velocity v with θ_{Av} being the dynamic advancing contact angle at that velocity. Rearrangement gives the dynamic contact angle as:

$$\theta_{Av} = \arccos \left(\cos \theta_{Av \to 0} - \frac{f_h}{\gamma_{vl}} \right) \tag{6.51}$$

Assuming the coating is a Newtonian fluid, the shear stress ($N\,m^{-2}$) is equal to the product of the viscosity and the shear rate. The latter is calculated from the tape velocity and the gap, z. So, we have:

$$\sigma = \eta \dot{\gamma} \text{ and } \dot{\gamma} = \frac{v}{z} \tag{6.52}$$

We need to characterize the area over which the stress is acting to give the force, and per unit length of wetting line we may use the distance into the liquid film where flow is occurring as nz, where $n \geq 1$. A value of $n \sim 1$ will simplify the calculations – this is a reasonable approximation – and so the viscous force contribution is:

$$f_h = \eta \frac{v}{z} nz \approx \eta v \tag{6.53}$$

The dynamic advancing contact angle is now:

$$\theta_{Av} \approx \arccos \left(\cos \theta_{Av \to 0} - \frac{\eta v}{\gamma_{vl}} \right) \tag{6.54}$$

Note that for the equivalent relationship for a receding angle the viscous contribution must be added:

$$\theta_{Rv} \approx \arccos \left(\cos \theta_{Rv \to 0} + \frac{\eta v}{\gamma_{vl}} \right) \tag{6.55}$$

Let us take an example to illustrate the process. If the static contact angle is $60°$ (no hysteresis), the surface tension of the coating is $50\,mN\,m^{-1}$ and the

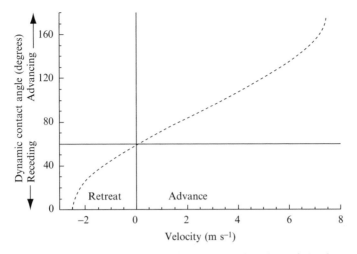

Figure 6.13. The dynamic contact angle plotted as a function of the front velocity, where the calculations were carried out for a solution with a static surface tension of $50\,mN\,m^{-1}$ and a viscosity of $0.1\,Pa\,s$.

viscosity is $0.1\,N\,m^{-2}\,s$ ($0.1\,Pa\,s$), we have the effect of tape velocity shown in Figure 6.13. We see from this figure that with this modelled coating system operating at a velocity of $\sim 7.5\,m\,s^{-1}$, the dynamic advancing contact angle is $\sim 180°$. When this occurs, air is entrained and the resulting coating is poor with patches of bare surface. The implication for the receding angle curve is that 'de-wetting' becomes ineffective at velocities $\sim 2.5\,m\,s^{-1}$ with such a model. This would be important in situations such as drawing a fibre out of a coating fluid. Even when a surface is completely wetted by the coating, air entrainment still occurs if the velocity is high enough. The results of an example calculation, which uses the same simple model, is shown in Figure 6.14. In this case, the entrainment is predicted at a quite low velocity of $\sim 50\,cm\,s^{-1}$ as the viscosity is much higher at $0.25\,Pa\,s$.

Many coatings consist of concentrated dispersions and the viscosity deformation rate curve is non-linear with the material showing *shear thinning*. The shape of the curve will be modified with a more rapid rise at low velocities as the stress is proportionately larger. The curve shape will also depend on the precise flow behaviour in the coating system and the model given above would need to be modified for the particular flow pattern and the change in that flow with coating velocity. However, it does illustrate the general behaviour quite adequately.

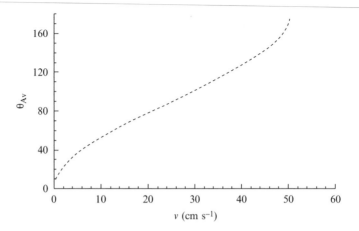

Figure 6.14. The dynamic contact angle calculated for a moving surface with a static contact angle of zero, a surface tension of $63\,\text{mN}\,\text{m}^{-1}$ and a viscosity of $0.25\,\text{Pa}\,\text{s}$. Air entrainment occurs as the velocity reaches $50\,\text{cm}\,\text{s}^{-1}$.

REFERENCES

1. A. W. Adamson, *Physical Chemistry of Surfaces*, 5th Edn, Wiley-Interscience, New York, 1990.
2. T. D. Blake, 'Wetting', in *Surfactants*, Th. F. Tadros (Ed.), Academic Press, London, 1981, pp. 221–275.
3. J. W. Gibbs, *The Collected Works of J. Willard Gibbs*, Vol. 1, Longmans, New York, 1928.
4. J. N. Isrealachvili, *Intermolecular and Surface Forces*, 2nd Edn, Academic Press, San Diego, CA, USA, 1991.
5. R. N. Wenzel, *Ind. Eng. Chem.*, **28**, 988 (1936).
6. L. A. Girifalco and R. J. Good, *J. Phys. Chem.*, **61**, 904 (1957).
7. F. M. Fowkes, *Ind. Eng. Chem.*, **56**, 40 (1964).
8. W. A. Zisman, *Adv. Chem. Ser.*, **43**, 1 (1964).
9. R. J. Hunter, *Foundations of Colloid Science*, Vol. I, Oxford University Press, Oxford, UK, 1987.
10. W. D. Harkins and G. Jura, *J. Am. Chem. Soc.*, **66**, 1362 (1944).

Chapter 7

Emulsions and Microemulsions

1 INTRODUCTION

Dispersions of one liquid in another imiscible liquid are of great importance in many applications, ranging from through such diverse fields as foods and pharmaceutics to the oil industry. Many of the general features of colloidal dispersions are also common to emulsions, such as the stabilizing mechanisms that counteract the attraction due to dispersion forces or the tendency of large particles to settle or 'cream' due to the effects of gravity. The fluid nature of both the particle and the interface results in special features that we do not observe with solid particles, although some of these are shared with the disperse phases in foams. Not everything that is commonly referred to as an emulsion is, in fact, a liquid-in-liquid dispersion. For example, the photographic 'emulsion' which is coated onto a film matrix is a dispersion of solid particles (silver halide and dye crystals depending on the film type) and has a fluid continuous phase when coated, but neither phase is a liquid when in the camera. Emulsion polymers used in decorative paints provide another example of this. The monomers are dispersed as an emulsion prior to polymerization, but once polymerized the particles are solid although their glass transition temperature is well below room temperature so that they can fuse to form a film. The polymer particles though never completely lose their individual identity and dissolve in one another.

Most common emulsions have a particle size distribution at the upper end of the colloidal size range, that is, with a radius in the 1 to 10 μm range. The problem here is that in their formation a great deal of energy is required to create the large amount of new interface that is produced with particles with radii of one or two orders of magnitude smaller. By using a mixed surfactant system, say an ionic surfactant with a co-surfactant such as an alcohol, droplets with a radius of ~ 100 nm can be produced and such systems are

Colloids and Interfaces with Surfactants and Polymers – An Introduction   J. W. Goodwin
© 2004 John Wiley & Sons, Ltd   ISBN: 0-470-84142-7 (HB)   ISBN: 0-470-84143-5 (PB)

frequently referred to a 'mini-emulsions'. Just like their larger brethren, these systems are inherently colloidally unstable. However, if the surfactant and co-surfactant contents are increased to high levels, a thermodynamically stable hetero-phase system may be formed with droplet radii in the 10 nm range. These systems are called 'microemulsions' and are a quite different type of colloidal dispersion from the mini- and normal emulsions. So, although both types are discussed in this chapter, they will be considered quite separately.

Many of the considerations that we apply to dispersions of one liquid in another apply to the dispersion of any fluid in a liquid, e.g. dispersions of a gas in a liquid are foams and are very similar to emulsions. Not all foams are of the very high-phase volumes that we are familiar with in the foams formed by the detergents used in the hand-washing of dishes or in shampoos. A low-phase-volume foam is termed a 'low-quality' foam, while a high-phase-volume foam (say at $\varphi > 0.75$) is termed a 'high-quality' foam. When we are referring to a liquid dispersed phase, we would use the terms 'emulsion' and 'cream' for the equivalent concentrations. The compressibility of the gas phase is much greater than that of a liquid dispersed phase and the solubility is also invari-ably much greater. Both of these factors provide additional layers of complex-ity. We will, however, not deal with foams specifically in this text but should keep them in mind when we think of emulsions.

2   EMULSIFICATION

When we make an emulsion, we disperse one liquid phase in a finely divided form in another liquid phase which is imiscible with the first. Hence, we may disperse water in an oil (a W/O emulsion), oil in an aqueous phase (an O/W emulsion) or one oil in another (an O/O emulsion). The latter is not very common but W/O and O/W emulsions are very widely used. It is also possible to turn one into the other – a process known as *phase-inversion*. For example, milk is an O/W emulsion which is phase-inverted to give a W/O emulsion which we recognize as butter. However, the first question that we need to address is 'how do we disperse one system in the other?'. During the emulsifi-cation process, we have to create a large amount of new interface. The free energy change is estimated from the product of the interfacial tension and the areal change. There is also a positive entropy change as we are dispersing one phase, and so we may write:

$$\Delta G_{\text{formation}} = \gamma_{12}\Delta A - T\Delta S \qquad (7.1)$$

where γ_{12} refers to the interfacial tension of liquid '1' against liquid '2', with the latter being the continuous phase. We are not dispersing the disperse phase into very small drops and so the entropy change is small:

$$\gamma_{12}\Delta A >> T\Delta S \tag{7.2}$$

This informs us immediately that emulsification is not a spontaneous process and that we have to provide the energy input. In part, this may be in terms of heating but it is mainly in terms of mechanical energy by using very high shear rates. Equation (7.1) tells us that the resulting droplets would be unstable, so that we must find a means of preventing them reforming into two distinct phases. The free energy change only tells us about the difference in the initial and final states, and so once formed we may provide a kinetic stability via the forces of repulsion that we employ with rigid particles. However, the details of the route is important in the formation process as well. When shear forces are produced by intense flow fields, the droplets are progressively broken down from large macroscopic drops to small drops in the colloidal domain. In order to disrupt a droplet, a force has to be applied to it, hydrodynamically in the case we are considering here. The pressure gradient must be large enough to exceed that due to the Laplace pressure:

$$\frac{dP}{dr} \geq \frac{\Delta P}{a} = \frac{2\gamma_{12}}{a^2} \tag{7.3}$$

Of course, the area produced by dispersing unit volume of phase 1 in phase 2 is simply $3/a$ as the droplets are spherical, and so we readily note by comparing Equation (7.1) with Equation (7.3) that the latter is where most of our work will come from.

The work required to break up the droplets may be applied in various ways. In a high-speed disperser or blender, large shear and elongational forces are applied to the droplets. The droplets elongate and break in the middle. As they separate, a small or 'satellite' drop is formed as the neck breaks. The newly formed drops must be rapidly stabilized as they tend to coalesce during subsequent collisions. When very high flow velocities are used, turbulent flow is usually produced. In turbulence, the flow is chaotic with localized eddies with high energy dissipation. The inertial forces are large and are the cause of the instabilities in the flow field, and hence the density of the continuous phase is important, with higher densities leading to turbulence at lower flow rates. The size of the local eddies can be defined in terms of the *Kolmogorov scale* (from Kolmogorov's analysis) and so we may write the droplet size as follows [1]:

$$a \sim \dot{E}^{-0.4}\gamma_{12}^{0.6}\rho_2^{-0.2} \tag{7.4}$$

where $\dot{E} = dE/dt$, the rate of energy dissipation per unit volume. An ultrasonic probe may also be used to produce emulsions. In this case, the local cavitation is used to break up the droplets.

In all cases though, the processes are not very efficient and there is normally significant heating[2]. This can aid the break-up as the timescale of the disruption process is also important. The viscosities of both phases decrease with increasing temperature and an important factor is the timescale of the extension of the droplets during break-up (slower with higher viscosity) compared to the timescale associated with the flow field (inversely proportional to the flow rate). The result is that larger drops are produced from disperse phases with higher viscosities. Of course, a higher viscosity of the continuous phase means higher shear forces on the drops and so a faster deformation, thus leading to a smaller drop size. However, the turbulence will not be as marked, with the flow having a larger eddy size, and so the beneficial effect of increasing the continuous phase viscosity may by reduced somewhat.

2.1 Emulsion Type

The control of which phase is to be the disperse phase is of major importance. The major controlling factors are the type of surfactant that is used and the temperature of the emulsification process, although the volume ratio and the viscosity ratio of the two phases can be important. The higher-viscosity phase tends to be the continuous phase, with the stability of the dispersed phase being enhanced due to the slower drainage of the thin film produced as two drops collide. The choice of surfactant is normally the controlling factor. There may be limitations when the emulsion is intended for a specific use, such as food or personal-care products. However, there are some general guidelines which can be useful.

The simplest of these is known as *Bancroft's rule* which states that the continuous phase should be the phase in which the surfactant is the most soluble. A good example with which to illustrate this is the water/cyclohexane system studied by Shinoda and co-workers[3, 4], in which a non-ionic surfactant (a nonylbenzene ethoxylate at 5 wt%) was used. We will consider what these authors observed at a phase ratio of 1:1. At room temperature, the surfactant is below its *cloud point* and is most soluble in the water phase, with the ethylene oxide groups strongly hydrated and fitting into the H-bonding structure of the water. The emulsification process results in an O/W emulsion being formed. At 80 °C, the surfactant is more soluble in the cyclohexane where it forms inverted micelles with the ethylene oxide groups oriented towards the centre of the micelles. This occurs as the hydration of the ethylene oxide moieties is reduced. An interesting question to address is 'what happens when an emulsion is formed at, say room temperature, and it is then heated to 80 °C?'. The result is that the emulsion can invert and the temperature at which this occurs is known as the *phase-inversion temperature* (PIT). In this case, the PIT was ~ 55 °C and in this region the surfactant produced a concentrated lamellar surfactant phase as the solubility was not high in either phase, with some cyclohexane droplets as well as some water droplets. The PITs occur at the cloud points of non-ionic surfactants in water, while with ionic surfactants they can occur at the Krafft points. The phase-inversion can be followed by observing changes in the physical properties of the system, such as conductivity or viscosity, for example.

Attempts have been made to classify surfactants numerically in terms of their chemical structure so that the selection of emulsifiers would be more straightforward. This was first attempted by Griffin[5] for non-ionic surfactants and the concepts extended by Davies and Rideal[6]. The *hydrophilic–lipophilic balance* (HLB) number is defined in terms of numerical values assigned to the chemical groupings in the surfactant, as follows:

$$HLB = 7 + \sum (\text{hydrophilic group numbers}) - \sum (\text{lipophilic group numbers})$$

$$(7.5)$$

The group numbers assigned by Davies and Rideal[6] are given in Table 7.1. It is interesting to note that while the ethylene oxide group is hydrophilic, the propylene oxide group is hydrophobic. This is confirmed by the effectiveness of the ABA block copolymers, PEO–PPO–PEO, to act as stabilizers. Thus, we may estimate the HLB numbers for our surfactant systems from Equation (7.5), for example:

Table 7.1. HLB numbers for various chemical groups, with data taken from Davies and Rideal[6]

Type	Chemical group	Group number[a]
Lypophilic	–CH–	0.475
	=CH–	0.475
	–CH$_2$–	0.475
	CH$_3$–	0.475
Hydrophilic	–SO$_4$Na	38.7
	–COOK	21.1
	–COONa	19.1
	–SO$_3$Na	11.0
	=N–	9.4
	Ester (sorbitan ring)	6.8
	Ester (free)	2.4
	–COOH	2.1
	–OH (free)	1.9
	–O–	1.3
	–OH (sorbitan ring)	0.5

[a]Examples of *derived* group numbers: ethylene oxide $(-CH_2 - CH_2 - O-) = 1.3 - 2(0.475) = 0.35$ – *hydrophilic*; propylene oxide $(-CH_2 - CH(CH_3) - O-) = 1.3 - 3(0.475) = -0.125$ – *lipophilic*.

Table 7.2. Classification of surfactant application by using the HLB range of the surfactant, according to Griffin[5]. Note the original classification was for non-ionic systems

Application	HLB range
W/O emulsifier	3–6
Wetting agent	7–9
O/W emulsifier	8–18+
Detergent	13–16
Solubilizer	15–18

- sodium dodecyl sulfate, $CH_3(CH_2)_{11}OSO_3Na$, has an HLB number of 40
- dodecyl hexaethylene gylcol monoether, $CH_3(CH_2)_{11}O(CH_2CH_2O)_5CH_2CH_2OH$ (or $C_{12}E_6$), has an HLB number of 5.3
- glycerol monostearate, $HOCH_2CH(OH)CH_2OOC(CH_2)_{16}CH_3$, has an HLB number of 3.7

When two surfactants are mixed together, the HLB of the mixture may be estimated from the HLB number of each component multiplied by the mass fraction of the component.

Griffin[5] used the HLB numbers to classify surfactants for particular uses and his classification is given in Table 7.2. Therefore, glycerol monostearate would be a suitable choice to produce a W/O emulsion while sodium dodecyl sulfate could be used to produce an O/W emulsion.

The HLB classification is a useful initial guide but is only an indication. A better solution is to use the *Hildebrand solubility parameter*. This can be related to the van der Waals forces between the components[7]; however, the HLB numbers are widely used. We should also note that the HLB values for non-ionic surfactants correlate well with the cloud point temperatures of such surfactants and so, in turn, should provide a guide to the values of the PITs.

3 STABILITY OF EMULSIONS

The aggregation of emulsion droplets is a function of the same parameters as were discussed for the aggregation of solid particles. Hence, it is important to know the value of the combined Hamaker constant, the magnitude of any electrical charge and the state of steric stabilizer layers. The fluid nature of the particle adds complexity as the interface can deform under the influence of both attractive forces and shear. This means that factors such as the density of the packing of steric layers or charges are slightly variable but also that the location of the stabilizing moieties can also rapidly change as lateral diffusion in the stabilizer layer can be rapid.

There are further additional factors that we need to consider. Perhaps the most important of these is that aggregation is frequently not the final state. The coalescence of particles to form larger particles, and finally two distinct phases, leads to a larger decrease in the free energy than is the case with aggregation alone. Of course, aggregation does not always have to lead to coalescence and stable dispersions of fluid particles and we will have to consider the details of the trilayer film formed by the two-droplet surface films separated by a thin film of the liquid disperse phase. Although two-body interactions can occur during collisions as the result of Brownian motion or collisions due to a shear field, we will usually have coalescence occurring in the concentrated state. This may be produced by aggregation and/or by gravitational separation. The latter is just sedimentation or creaming if the disperse phase is less dense than the continuous phase. Monodisperse emulsions are very unusual and the wide particles size distributions of most emulsions mean that droplets at $\varphi < 0.75$ are spherical, albeit with flattened interaction zones if in the aggregated state. Above this concentration, we have creams and the droplets become increasingly deformed as the dispersed-phase volume is increased. The smaller droplets are more rigid than the larger ones and the latter deform first, giving polyhedral shapes with some surfaces concave due to the presence of particles with smaller radii of curvature. This is illustrated in Figure 7.1(a).

Many common emulsions have the majority of their mass of material in droplets with radii of curvature in the 0.5–110 µm range. It is relatively easy to deform droplets of these dimensions and so there are flat parallel areas between the particles. The interparticle forces acting across the thin films determine the mean separation distance. These are the dispersion forces, electrostatic forces and/or steric forces discussed in detail in Chapters 3 and 4. Such forces are formulated from static or equilibrium models, but when considering the stability to coalescence of emulsions (and many of the points are also directly applicable to foams), the dynamics are a major factor. When two fluid droplets come together, whether during a Brownian collision, a

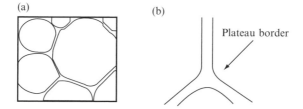

Figure 7.1. Schematics of (a) a section of a concentrated emulsion illustrating the polyhedral shapes of the droplets, and (b) the junction between three droplets, showing the thin-film region with parallel faces and the region of varying curvature known as the *plateau border*.

shear-induced collision or due to the gravitational force, the following processes occur. The rate of close approach is determined by the balance of those forces bringing the particles together, the interparticle forces, and the viscous force from the fluid drainage as we would anticipate for rigid particles. In addition, we have to consider the changes in the interaction region, i.e. at close separations the interaction zones flatten and this increases both the strength of the interactions and the viscous resistance to the drainage as the potential contact area is increased. As the interaction zone grows, the area of the interface increases. The rapid stretching of the interface results in a locally higher interfacial tension and this opposes the stretching. The difference in the dynamic surface tension and the static value resists the expansion and acts as a driving force to move the adsorbed stabilizing material into the interaction zone. This flow also causes fluid to be drawn into the thinning film region by viscous drag and, in addition, opposes the drainage. These effects are grouped together by the term *Marangoni effect*. There is a further effect, known as the *Gibbs effect*, which is governed by the decrease in adsorbed amount of stabilizer at the expanded interface which increases the stiffness (dilational elastic modulus) of the film. In thin films, the local mass concentration of stabilizer can be too low to re-establish the initial equilibrium. These effects are usually grouped together and referred to as the *Gibbs–Marangoni effect* and act to resist deformation of the interfaces in the thin-film region. However, we must also think in terms of local thermally (diffusion) driven 'ripple-like' disturbances. The Gibbs–Marangoni effect will act to damp these out but they are a main mechanism by which coalescence occurs. To understand this, we must visualize the thin intervening film between two fluid particles where local wave effects cause local fluctuations in the separation. The stiffer the film, then the smaller are the fluctuations. The force which maintains the film, known as the *disjoining pressure*, is the sum of the attractive forces and the repulsive forces. The elastic modulus is the rate of change of the force with distance and is again the sum of the attractive and repulsive terms. We may write this as follows:

$$\text{modulus} = \frac{\partial F_R}{\partial H} - \frac{\partial F_A}{\partial H} \tag{7.6}$$

It is quite clear from this equation that if $\partial F_R/\partial H < \partial F_A/\partial H$ the stiffness decreases and the magnitudes of the 'ripples' (local thinning) increase and coalescence will occur.

We can summarize the factors that lead to a stable emulsion as follows:

(1) The stabilizer on the surface of the droplets should produce sufficient repulsion to maintain a film of the continuous phase between droplet surfaces during collision or in a concentrated system such as a sediment/cream.

(2) The stabilizer should be strongly adsorbed at the interface so that it will not desorb as droplets come together.
(3) The dilational modulus of the stabilizer layer is high so that the thermal disturbances are damped and will grow to a magnitude comparable with the dimensions between adjacent drop surfaces.
(4) The solubility of the disperse phase in the continuous phase should be as low as possible in order to minimize diffusive transport from one droplet to another in the process known as 'Ostwald ripening'.

3.1 Ostwald Ripening of Emulsions

This is the expression given to the process whereby the droplet size distribution in an emulsion progressively shifts towards larger sizes. The origin of the effect is the Laplace pressure acting to increase the pressure inside the droplet. As a result of the pressure increase, molecules of the dispersed phase diffuse from the high-pressure regions to the low-pressure ones, that is, the small droplets dissolve and the larger ones grow as the material is transferred by diffusion through the continuous phase.

The Laplace pressure is given by Equation (6.24) (see earlier) for droplets in a vapour, and for a drop of radius r_1 in a liquid, we have:

$$P_{r_1} = P_{r=\infty} + \frac{2\gamma_{12}}{r_1} \qquad (7.7)$$

where γ_{12} is the interfacial tension. Clearly, if there are two droplets with different radii the difference in pressure between them is:

$$P_{r_2} - P_{r_1} = 2\gamma_{12}\left(\frac{1}{r_2} - \frac{1}{r_1}\right) \qquad (7.8)$$

and this acts to equalize the radii at a rate governed by the diffusion of the dispersed phase through the continuous phase. In Chapter 6, we also discussed the Kelvin equation (Equation (6.40)), which describes the change in vapour pressure with drop radius. The analogous situation for a drop in a liquid (or a solid particle in a liquid, for that matter) is the solubility of the dispersed phase in the continuous. So, if c_r is the concentration of the dispersed phase in the continuous phase as a result of the applied pressure P_r, we can write the analogue of Equation (6.40) for the solubility of the dispersed phase as follows:

$$\ln\left(\frac{c_{r_1}}{c_{r=\infty}}\right) = \frac{\overline{V}_1}{RT}\frac{2\gamma_{12}}{r_1} \qquad (7.9)$$

and for two drops we have therefore:

$$\ln\left(\frac{c_{r_1}}{c_{r_2}}\right) = \frac{M_1}{\rho_1 RT} 2\gamma_{12}\left(\frac{1}{r_2} - \frac{1}{r_1}\right) \tag{7.10}$$

where M_1 and ρ_1 are the molar mass and density of the dispersed phase, respectively. Hence, Ostwald ripening is most marked if the distribution of drop sizes is broad and the interfacial tension is high.

The data [8] shown in Figure 7.2 illustrate the relative growth in the mean particle size of an oil-in-water emulsion after a period of 10 d. The stabilizer was sodium dodecyl sulfate and the initial mean droplet size was ∼ 1 μm. This figure also clearly shows how the relative growth rate follows the solubility of the alkane in water.

The prevention of Ostwald ripening is usually desirable as ageing a formulation is not an attractive proposition. The solubility of the dispersed phase in the continuous phase is a factor which effects the rate rather than the ultimate state. The interfacial tension may be manipulated somewhat but this is also likely to change the initial size distribution. The addition of small amounts of a third component, which is soluble in the dispersed phase but has a extremely low solubility in the continuous phase, is an effective strategy to limit the size drift. In this case, the concentration of the third component is the same in all of the particles after emulsification. As ripening commences, the concentration in the small particles increases and that in the large ones decreases and this opposes and ultimately limits the ripening process. (The situation in the droplets is directly analogous to an osmometer where the continuous phase is acting as the semi-permeable membrane.) Now, if the surfactant has a very low solubility in the continuous phase while being

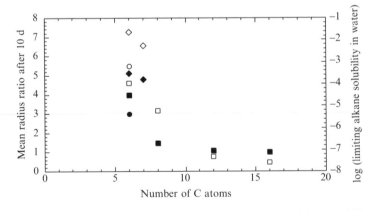

Figure 7.2. The Ostwald ripening behaviour of oil/water emulsions stabilized with sodium dodecyl sulfate, taken from Winsor [8]: ■, □, *n*-alkanes; ●, ○, cyclohexane; ◆, ◇, aromatic alkanes (where the solid symbols refer to data for the radius ratios and the open symbols to the alkane solubilities).

sufficiently surface active to act as a stabilizer, of course, it will act as this third component. The reduction in the specific surface area that occurs as the ripening proceeds increases the concentration of surfactant dissolved in the small droplets. It is a common practice to dissolve surfactants in what will become the dispersed phase.

4 MICROEMULSIONS

The interfacial tension is an important parameter in the control of droplet size in emulsification. This may be reduced by the addition of co-surfactants such as long-chain alcohols. Droplet sizes of $\sim 100\,nm$ may be produced in this way and the resulting emulsions are referred to as 'mini-emulsions'. These differ little from emulsions with larger droplet sizes in terms of stability, although the rate of creaming/sedimentation is reduced and this may no longer be a problem in a formulated product. If interfacial tensions are reduced to very low levels though, a further reduction in droplet size can occur down to $\sim 10\,nm$. The systems are produced at moderate-to-high levels of surfactant and are transparent due to the small size of the dispersed phase. The viscosity is usually low, unlike liquid crystal phases, and the stability is quite different from what we regard as 'normal emulsions'. Such systems represent thermodynamically stable phases and are termed *microemulsions*. Formation is spontaneous, requiring little or no input of mechanical energy beyond a gentle mixing of the components. This inherent stability means that the systems are quite different from 'normal emulsions' and have very little in common with them except that they are colloidal systems consisting of oil, water and surfactants.

Microemulsions have been used in many formulations[9, 10]. Cleaning systems range from dry-cleaning fluids to self-polishing floor and car waxes. The hard wax ('Carnuba' wax or other synthetic polymers) is in the oil phase, and when the coating dries the wax particles form the film. As the dimensions are so much smaller than the wavelength of light, the surface asperities are too small to result in individual reflections. Other examples of applications are in cutting oils, pesticides and flavours for foods. There are some examples of chemical reactions being carried out in microemulsions, with the most successful of these being the polymerization of acrylamides to give very high-molecular-weight products[11]. An interest in the use of microemulsions in tertiary oil recovery increases and decreases in parallel to oil prices. The attraction is the spontaneous formation of the microemulsion so that oil trapped in the pores of depleted oil wells (although there can be 75 % of the original amount remaining) can be flushed out. The large amounts of surfactant required is the cost-limiting factor. Microemulsion fuels using vegetable oils from sustainable sources have also been formulated[10].

4.1 Phase Behaviour

The phase diagrams for microemulsions are quite complex as we have at least three components (and often four), namely the oil, water (electrolyte solution) and surfactant (and sometimes a co-surfactant). We must also remember that temperature and pressure are going to be important in many situations. A diagrammatic representation of a six-component phase map is not possible – just four requires a tetrahedron but the usual practice is to reduce the representation to a three-component map. This may be achieved by working at a constant surfactant/co-surfactant mix, in addition to constant temperature and pressure. Alternatively, the surfactant concentration in the electrolyte solution can be kept constant and the co-surfactant concentration can be the third variable. This enables us to utilize the conventional triangular phase-diagram plot. Temperature variations are then represented by slices across a parallel-sided triangular prism. The plotting of the phase diagrams is a lengthy endeavour as many samples are required to precisely define the boundaries. The process may be speeded up by using a titration technique [8], where two of the components are adjusted sequentially. A schematic of a typical phase map is given in Figure 7.3.

A wide range of surfactants can produce microemulsions. Anionic surfactants, such as sodium dodecyl sulfate or potassium oleate, require co-surfactants such as aliphatic alcohols of shorter chain length. As an alternative to the alcohols, some non-ionic surfactants can be used in which a polyethylene glycol moiety replaces the simple alcohol, although it should be noted that the properties are a little different [10]. Some di-chain ionic surfactants, such as Aerosol OT (sodium di-(2-ethylhexyl)sulfosuccinate) and didocecylammonium bromide, do not require a co-surfactant to form microemulsions with oil and water. The same is true for some non-ionic surfactants. This occurs in the phase inversion boundary region between W/O and O/W systems.

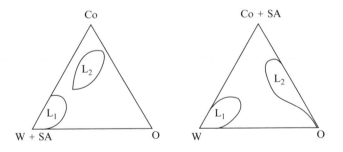

Figure 7.3. Schematic ternary phase diagrams showing the oil-in-water microemulsion region, L_1, and the water-in-oil microemulsion region, L_2: O, oil; W, water; SA, surfactant; Co, co-surfactant.

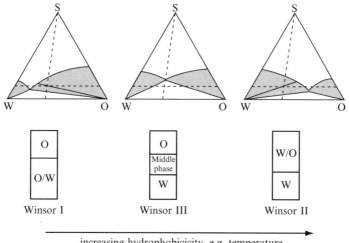

Figure 7.4. Schematic phase diagrams of various types of microemulsion systems, as classified by Winsor[8]: O, oil; W, water; S, surfactant. The overall compositions are given by the intersections of the dashed lines, the shaded areas represent the microemulsion plus pure component regions, while the ratios of the microemulsion phases and the pure liquid phases are given by the 'tie-lines' and the 'lever rule'.

Figure 7.4 shows schematically the type of microemulsion systems that were classified by Winsor[8], where we see microemulsion phases in equilibrium with excess water or oil. The composition of the three-component systems are shown as the intersections of the broken lines. The ends of the 'tie-lines' give the compositions of the two phases, while the 'lever rule' gives the amounts of each. Any change to the system which changes the hydrophobic balance of the surfactant may enable us to move from one type of system to another. For example, at a constant temperature the change of a surfactant to one with a longer chain length would take us from left to right in the figure. A similar result can be achieved by changing the polarity of the water phase (by the addition of short-chain alcohols, for example), increasing the temperature for a non-ionic surfactant or increasing the electrolyte concentration for an ionic species. This is the same process as described above for the phase-inversion temperature for the non-ionic system studied by Shinoda and Kunieda[4] where the H-bonding in the water is reduced. This decreases the solubility of the hydrophile moiety and increases the solubility of the hydrophobe. The head group of an ionic surfactant becomes less soluble with increasing added electrolyte. This is readily observed as the cmc decreases with added salt as the ion binding is increased and the mutual repulsion decreased (i.e. we are tending to 'salt-out' the molecule). So, just as for macroemulsions, we have the continuous phase as the one in which the surfactant is most soluble, as described by Bancroft's rule. The curvature of the surface is governed by the packing of the molecules at the interface and this, of course, includes solvation and electrostatic interactions. For the formation of such small droplets as we find in microemulsions (5–50 nm), the interfacial tension has to be 'ultra-low', with typical values being in the range from 10^{-4} up to 10^{-2} mN m^{-1}.

The Winsor III system (see Figure 7.4) is particularly interesting. The concentration of the system is outside the two-phase regions marked in the figure and is in a multi-phase region. The HLB of the surfactant is such that we have equal solubility and no preferred curvature. This means that the interfacial tension is not just 'ultra-low' but is approaching zero. The rigidity (elasticity) of the interface is therefore low and the interface is readily deformed by thermal fluctuations so that curvatures in favour of both the water and the oil directions are probable. A 'bicontinuous' or 'sponge' microemulsion phase is produced, with a domain size typical of the microemulsion size domain, that is, ~ 10 nm. If the rigidity of the film is increased, by the use of long-chain alcohols as co-surfactants, for example, the bicontinuous middle phase has a lamellar structure with oil swelling the hydrophobe region.

4.2   Characterization of Microemulsions

The first problem that we must solve is 'which type of system do we have?'. Conductivity is an easy check here. If the conductivity is low, the system is a water-in-oil microemulsion. Conversely, if it is high then we have an oil-in-water system. A bicontinuous phase would also have a high conductivity and we would need other information to separate this from the O/W microemulsion.

Transmission electron microscopy, coupled with a freeze-fracture technique, is especially useful in studying the bicontinuous phase [12]. In this, we use liquid nitrogen to very rapidly cool a small sample so fast that the water is turned to 'amorphous ice'. If we were to put a sample straight into liquid nitrogen, a gaseous layer around the sample would act as an insulator – therefore, we place the sample into a volume of liquid propane which is held at liquid nitrogen temperatures. The cooled sample is placed in a high vacuum, fractured and then coated with tungsten to give a shadow – a replica is next formed from a platinum/carbon coating, which can then be viewed in the electron microscope.

Dynamic light scattering can be used to follow the diffusive motion of the individual droplets. However, the samples are relatively concentrated and the diffusion is slowed by particle–particle interactions. NMR spectroscopy can be employed to determine the molecular diffusion – in this case, a technique known as pulsed-gradient spin-echo NMR spectroscopy is used.

Small-angle neutron scattering is a particularly powerful technique as the use of D_2O and deuterated oils and surfactants and/or co-surfactants enable detailed information of the droplet structures to be obtained. Small-angle X-ray scattering may also be useful but it is not possible to vary the contrast to bring out the detail with this technique, as is the case with neutron scattering. In addition, the scattering is much weaker when using X-rays as the elements are all of low atomic number and the X-rays are scattered by the electrons. Neutrons are scattered by the atomic nuclei and thus give stronger scattering.

4.3 Stability of Microemulsions

It was stated above that microemulsions are a *thermodynamically stable* phase, and are therefore unlike macroemulsions. The modelling of these systems is a more recent development than that of the stability of other colloidal systems and is still not fully complete. Most practical systems contain many components and usually these will contain impurities which will also be surface active. Nevertheless, it is instructive to consider the main ideas involved in modelling the stability, although we will not follow the detail of the analysis.

The foundations were laid in the work of Overbeek[13] and Ruckenstein and Krishnan[14]. These workers discussed the free energy change in the formation of a microemulsion consisting of droplets, although we will not broaden the discussion to the details of the bicontinuous or sponge phase. Following these authors, we may split the free energy of formation into several components:

$$\Delta G_{\text{disp}} = \Delta G_{\text{I}} + \Delta G_{\text{E}} - T\Delta S_{\text{disp}} \qquad (7.11)$$

where ΔG_{I} is the free energy associated with creating the interfacial area of the droplets. The electrostatic components associated with the interactions produced by charged surfactant molecules are represented by the ΔG_{E} term and are a function of the curvature, that is, droplet radius, and also of the relative interfacial concentrations of the charged surfactant and uncharged co-surfactant. The entropic term is the contribution of the fine droplets in a random dispersion. To obtain a thermodynamically stable system, the free energy of the microemulsion phase must be lower than that of the original mixture of macroscopic phases. Hence, the value of ΔG_{disp} must be negative. Being a little more specific, we can write:

$$\Delta G_{\text{I}} \propto \gamma_{\text{ow}} A \qquad (7.12)$$

where A is the area of the interface that has been created with an interfacial tension of γ_{ow} and this is positive. As the droplets are small, the value of A is necessarily large and so the interfacial tension has to be 'ultra-low'. The electrical work needed to get two drops to their mean position is calculated from the electrostatic potential multiplied by the charge:

$$\Delta G_E \propto \int \psi d\sigma \qquad (7.13)$$

This again is positive as particles of similar charge repel and confining many charged particles into the space available to the macroscopic phases will require the input of energy. The entropic term for n particles occupying a space is given by $nk_B T$ and so:

$$T\Delta S_{disp} \propto k_B T \ln\left(\frac{\varphi}{\varphi_m}\right) \qquad (7.14)$$

The volume fraction, φ, is that of the disperse phase, while the maximum volume fraction that could be occupied by hard spheres is given by φ_m. The droplets in a microemulsion are spherical and monodisperse and in a static system we could imagine that we might pack them together in either a face-centred cubic or hexagonal close-packed structure which would give a maximum volume fraction of 0.74. The dense random packing of hard spheres is lower, at a maximum volume fraction of ~ 0.62. The thermal motion of the droplets means that each has a larger excluded volume that we might predict for a static system and a phase transition occurs at a volume fraction of $>$ 0.5, and even lower when particles are charged (this point is discussed in more detail in Chapter 9). We can see immediately from Equation (7.14) that for $\varphi < \varphi_m$ the entropic term is negative and for a stable microemulsion system to be formed, this term must be larger than those given by Equations (7.12) and (7.13). The work carried out by Ruckenstein and Krishnan[14] included adsorption isotherms of the surfactant and co-surfactant at the interface and this modifies both the interfacial tension and charge terms.

Although this discussion has referred to 'droplets' and 'hard spheres', it is important to keep in mind that the radius of a microemulsion droplet is of the same order of magnitude as the length of the surfactants used, noting that any co-surfactants are invariably shorter molecules. So, as the droplet size is on the same scale as the molecules, the interfacial structure is much less sharp, with significant penetration of the two phases into the interfacial region. The molecular shapes and the packing together of these shapes in a curved surface [7, 10, 15] is a critical consideration. In an O/W microemulsion, the surfactant chains are on the inside of the droplet interfacial region and are much more crowded than in a W/O microemulsion formed from the same surfactant system. When we refer to the diameter of a droplet, this is just some average value and different methods of determination will give different values. It is instructive to follow the estimate of water core size given by Overbeek *et al.* [15] for the water (300 mM NaCl)/cyclohexane/sodium dodecyl sulfate (SDS)/

co-20 % pentanol system. These workers used the limiting slope of the interfacial tension versus the log (surfactant concentration) curve to give the surface excess from the Gibbs adsorption equation (see Chapter 1). Their results gave one SDS molecule for every 0.9 nm², so that each of these molecules would have three pentanol molecules 'associated' with it in the macroscopic interface, which is, of course, flat. Assuming that this molecular area is also that at the curved surface of the water core of the droplet, we can calculate the water droplet surface area, A_D (in m²), for the number of moles of surfactant in the system, n_{SDS}, as follows:

$$A_D = 9 \times 10^{-19}(n_{SDS}N_{Av}) \approx 5.4 \times 10^5 n_{SDS} \qquad (7.15)$$

As the molar volume of water is 1.8×10^{-5} kg m^{-3}, the volume of water in the system is:

$$V_{H_2O} = 1.8 \times 10^{-5} n_{H_2O} \qquad (7.16)$$

Recalling that for a sphere, $V/A = (\pi/6)D^3/\pi D^2 = D/6$, we can use Equations (7.15) and (7.16) to give the approximate diameter (in nm) of the water core as:

$$D_{H_2O} = \frac{6V_{H_2O}}{A_{H_2O}} \approx 0.2\left(\frac{n_{H_2O}}{n_{SDS}}\right) \qquad (7.17)$$

We can make a similar estimate for the O/W system, although in this case the SDS/pentanol ratio is 1:2 [10, 15]. The estimates of droplet size are quite close to those measured [15].

The measurement of the limiting slope of the γ_{OW} versus log c_{SA} curve is not easy as the interfacial tensions are very low. A useful technique to use is the 'spinning-drop' method. In this case, the denser phase (usually water) is placed in a capillary tube. The tube is rotated at high speed and a small amount of the less dense phase (usually the oil) is added. The centrifugal forces keep the droplet centred in the tube. These forces also elongate the drop. The axial ratio is measured by using a travelling microscope and its value is limited by the value of the interfacial tension and the angular velocity of the capillary. This technique is the only one readily available for the measurement of ultra-low interfacial tensions.

REFERENCES

1. E. Dickinson, *An Introduction to Food Emulsions*, Oxford University Press, Oxford, UK, 1992.

2. P. Walstra, 'Formation of Emulsions', in *Encyclopedia of Emulsion Technology*, Vol. II, P. Becher (Ed.), Marcel Dekker, New York, 1983, pp. 57–77.
3. H. Saito and K. Shinoda, *J. Colloid Interface Sci.*, **32**, 647 (1970).
4. K. Shinoda and H. Kunieda, *J. Colloid Interface Sci.*, **42**, 381 (1973).
5. W. C. Griffin, *J. Soc. Cosmet. Chem.*, **1**, 311 (1949).
6. J. T. Davies and E. K. Rideal, *Interfacial Phenomena*, Academic Press, London, 1963.
7. J. Isrealachvili, *Intermolecular and Surface Forces*, Academic Press, London, 1991.
8. P. A. Winsor, *Chem. Rev.*, **68**, 1 (1968).
9. L. M. Prince, *Microemulsions*, Academic Press, New York, 1977.
10. R. J. Hunter, *Foundations of Colloid Science*, Vol. II, Oxford University Press, Oxford, UK, 1989.
11. F. Candau, Y. S. Leong and R. M. Fitch, *J. Polym. Sci., Polym. Chem. Ed.*, **101**, 167 (1984).
12. W. Jahn and R. Strey, *J. Phys. Chem.*, **92**, 2294 (1988).
13. J. Th. G. Overbeek, *Discuss. Faraday Soc.*, **65**, 7 (1978).
14. E. Ruckenstein and R. Krishan *J. Colloid Interface Sci.*, **76**, 201 (1980).
15. J. Th. G. Overbeek, P. L. de Bruyn and F. Verhoeckx, 'Microemulsions', in *Surfactants*, Th. F. Tadros (Ed.), Academic Press, London, 1984, pp. 111–132.

Chapter 8

Characterization of Colloidal Particles

1 INTRODUCTION

It is a prerequisite to understanding the way that colloidal systems behave to be able to know the details of the particles in our dispersions, such as the particle size, the size distribution, the shape, the charge and the dimensions of any adsorbed layers. There is a wide range of methods available and we shall review some of them in this chapter. In many situations, we may be restricted to which ones can be applied as well as to only some of these techniques being readily available. However, it is always true that the better our system is characterized, then the better our understanding and the easier it is to manipulate the formulation to optimize the material.

In this chapter, we will discuss the use of microscopy in terms of size and shape distributions. Following this, we then address the scattering of radiation by colloidal particles. This will focus on particle size and structure although it sets the scene for the discussion of concentrated systems in the next chapter. Characterization of the electrical properties is discussed next and finally the viscosity of dispersions is addressed as this is a function of the detailed nature of the particles.

2 PARTICLE SIZE

The first question that we ask when we discuss a colloidal dispersion is 'What is the particle size?'. Of course, we mean the mean size and we would also like to know something about the size and shape distributions as well as the

Colloids and Interfaces with Surfactants and Polymers – An Introduction   J. W. Goodwin
© 2004 John Wiley & Sons, Ltd   ISBN: 0-470-84142-7 (HB)   ISBN: 0-470-84143-5 (PB)

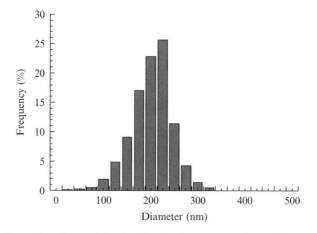

Figure 8.1. Example of a particle size distribution presented as a histogram: number-average diameter = 201 nm; standard deviation = 19 nm; coefficient of variation = 9.4 %.

specific surface area, but this latter information is often seen as of less import-ance than 'the size'. Before the methods are discussed, it is important to give some basic definitions.

An example size distribution is shown in Figure 8.1 as a histogram. The number average diameter, $\overline{D}_n$, is calculated in the same way as the average molecular weights for polymers, discussed in Chapter 2:

$$\overline{D}_n = \frac{\sum n_i D_i}{\sum n_i} = \sum f_i D_i \tag{8.1}$$

where n_i is the number and f_i the fraction of particles that are found in the size class D_i. We group similar sizes together to plot them as a histogram. Experimentally, of course, this grouping might conveniently correspond to the multiple of the resolution of the instrumentation. The standard deviation, σ_s, and the coefficient of variation, cv, are now:

$$\sigma_s = \left[\sum f_i (D_i - \overline{D}_n) \right]^{0.5}; \ cv = \frac{100\sigma_s}{\overline{D}_n} \tag{8.2}$$

We can calculate the various moments of the distribution, M_j, by using the following equation [1]:

$$M_j = \sum f_i D_i^j \tag{8.3}$$

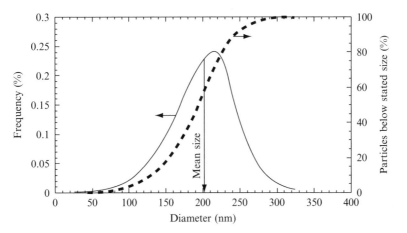

Figure 8.2. Frequency and cumulative distribution curves for a (colloidal) particle: number-average diameter = 201 nm; standard deviation = 19 nm; coefficient of variation = 9.4 %.

Thus, the first moment is the number average given in Equation (8.1) and the weight average is the ratio of the fourth to the third moment:

$$\frac{M_4}{M_3} = \frac{\sum D_i f_i D_i^3}{\sum f_i D_i^3}, \text{ as } w_i = \rho f_i D_i^3 \tag{8.4}$$

where w_i is the mass of particles with a diameter D_i and density ρ. In the example distributions plotted in Figures 8.1 and 8.2, the number-average diameter is 201 nm with a standard deviation of 19 nm, which is equal to a coefficient of variation of 9.4 %. The weight-average diameter is 226 nm. Here, we can see how the average changes as we use different weightings. We must keep in mind that size determinations which use different physical properties of the particles can correspond to different averages and so quite different sizes may be observed, especially if the size distribution is broad.

The shape of the distribution is indicative of how the system was produced. The example illustrated above (Figure 8.2) has a skew showing a tail towards smaller sizes. This is typical of the distribution that we find from a dispersion which is prepared by a particle nucleation and growth procedure. The tail indicates that new particles are formed until late in the process. If the system has been prepared by comminution, the tail is usually towards the larger sizes as we progressively reduce the size down to a limiting threshold. It is conventional to refer to a system as being 'monodisperse' if the coefficient of variation on the mean diameter is ≤ 10 %. When viewing a size distribution, especially when presented on a logarithmic scale, it is important to keep in

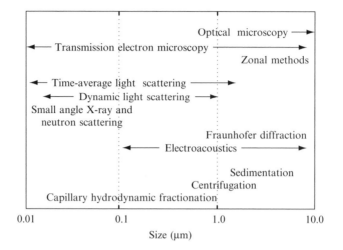

Figure 8.3. Summary of the techniques available for the measurement of particle size.

mind that a small size fraction can dominate the number and area of the system. In addition, it should be noted that the large size fraction of the distribution contains most of the mass of the disperse phase.

There is a wide range of techniques available for the determination of particle size. A summary of the more common methods is given in Figure 8.3. Some of these methods give the full distribution, while others provide just a mean value. The average is not always the number average and the measure of the distribution width is often determined by an algorithm which is embedded in the instrument software, the details of which are not available to the user.

3   MICROSCOPY

3.1   Optical Microscopy

The optical microscope is a valuable tool in any colloid laboratory. With modern optics, we can see much of the colloidal size range although we are still restricted to the larger end of the range for size data. The state of the dispersion is immediately apparent as it is observed in an 'unperturbed' state. Aerosols are an exception here as the motion is so rapid that high-speed photography would be required. With particles dispersed in a liquid, we can see immediately if the system is well dispersed, coagulated or weakly floccu-lated. In the latter case, particles can be seen to move together for some time and then to separate. The Brownian motion of particles dispersed in any

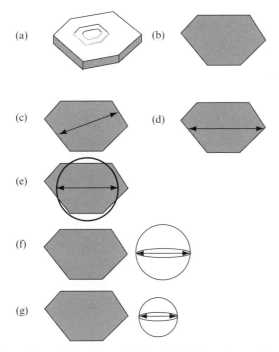

Figure 8.4. Illustration of the determination/definition of particle size for a non-spherical particle, in this case, a platelet: (a) three-dimensional 'image', showing the surface topography; (b) projected 'image', which loses the surface topography; (c) length bisecting the particle – Martin's diameter, d_M; (d) perpendicular distance between tangents of opposite sides – Feret's diameter, d_F; (e) diameter of circle of equal projected area, d_a; (f) diameter of a sphere with the same surface area, d_s; (g) diameter of a sphere with the same volume, d_v.

liquid presents problems with resolution if an accurate size is required. Dried samples may solve this problem but often a study of the particles in the wet state is important. A good photographic recording system is always required to optimize the method, whether normal transmission, dark field or fluorescence microscopy are employed.

Once an image has been obtained it has to be analysed by hand, or usually by a computerized image-analysis system. If the particles are spherical, the determination of particle size is straightforward – we simply define the diameter as the size. However, for non-spherical particles the problem is not so easy, as illustrated in Figure 8.4. In this figure, the particle is depicted as a platelet which would be similar to a kaolinite clay particle. The clay crystal has surface topographical detail which is lost in the projected image. The various dimensions are also defined in the figure (the dimensions shown in

Figure 8.4(f) and 8.4(g) would be obtained by techniques other that microscopy, of course, but are included here to illustrate the difficulties often encountered when 'sizes' are compared for the same material). Calibration requires an image for a stage micrometer, with the best precision being $\sim 3\%$.

3.2 Electron Microscopy

Transmission electron microscopy (TEM) is the preferred technique for size analysis if optimal precision is required. Here, a sample of the dispersion is placed on a mesh grid coated with a thin polymer or carbon film. When it has dried, the grid is placed in the microscope column which is then evacuated. The beam of electrons produces an image which is focused onto a fluorescent screen. The principle is exactly analogous to that of a 'normal' optical microscope. The lenses are electromagnets, and the wavelength of the electron beam is controlled by the accelerating voltage used in the microscope, with 10 or 20 kV being typical values, so that a resolution of < 1 nm is available. The limit though is not inherent in the instrument but more often is governed by the electron density of the particle and its thermal stability. Particles of high-atomic-number elements scatter electrons more strongly than low-atomic-number elements such as carbon. The heating effect of the beam is greatest in the highly focused intense beam used at the highest magnification and this can cause particles to shrink or decompose. The lowest magnification is limited by 'pincushion' distortion of the image which is readily observed on the screen. The reliable range for particle size analysis covers particles from ~ 10 nm to 10 μm. The magnification has to be calibrated with each sample and the standards used are crystal spacings for the higher magnifications and carbon replicas of gelatine casts, taken from diffraction gratings, for the lower magnifications. Like optical microscopy, a precision of 3 % is the best that we can usually achieve.

Scanning electron microscopy (SEM) is frequently available from the same instrument. In this mode, the electron beam is focused down to a spot of ~ 5 nm and scanned across the specimen. The latter is mounted on an aluminium stub and usually coated with a conducting material such as a thin film of gold. This prevents electrostatic charge building up which interferes with the image. The scanned beam produces low-energy secondary electrons from the sample surface which are collected and displayed on a TV monitor. The image that we see gives us a three-dimensional perspective of the sample surface which can be immensely valuable and is always of interest. However, in terms of size analysis, the 'foreshortening' results in a wider distribution of measured sizes than is actually the case. The resolution is less than with TEM and is limited by the size of the focused and scanned spot. If the conducting coating is absent, the energy distribution of the secondary electrons can be analysed and the composition of the surface layers can be determined. This technique is termed *electron probe microanalysis*.

In general, with electron microscopy we are limited to dry specimens as a high vacuum is required in order to prevent scattering of the electron beam by gas molecules. Microscopes are available in which the sample chamber is isolated from the rest of the microscope, except by a small aperture. The evaporation rate is slowed by the aperture and with a sufficiently powerful vacuum system a well-defined beam can be achieved. This enables wet samples to be imaged. However, the sample system is subject to evaporation and therefore is rapidly changing, which hence limits the utility of the technique.

3.3   Atomic Force Microscopy

Atomic force microscopy (AFM) is also a scanning technique. In this approach, a fine stylus mounted on a delicate cantilever spring is brought into close proximity with the surface being studied. The surface is displaced backwards and forwards until the selected area has been examined. The displacement of the stylus is monitored by using a laser beam and an image is built up on a computer monitor. A group of piezoelectric crystals are used to provide the fine three-dimensional displacement of the specimen. The optical lever and the spring constant control the resolution available, with this being similar in magnitude to that achieved by using SEM. It is relatively easy to use this technique with wet surfaces without any of the complications of electron microscopy.

With colloidal systems, the most important application of AFM has been in the study of (colloidal) particle interactions. A polystyrene latex particle is a convenient model, spherical, colloidal particle and, by attaching a latex particle to the stylus with cyanoacrylate adhesive, the displacement of the stylus as it is brought into close proximity to another particle is monitored. From a knowledge of the spring constant and the displacement, the force–distance curve can be constructed. The limitation of this approach is that the particles must be large enough to be attached to the stylus without the material of the latter playing a part in the interaction. Particles greater than 10^3 nm have been examined successfully with this technique.

4   ZONAL METHODS

In these methods, a dilute colloidal dispersion is caused to flow through a well-defined zone and changes in either the electrical or optical properties of the zone are monitored. The magnitude of the change is related to the size of the particles and the number is given by a direct count. A major problem is that we must take care to avoid the coincidence that two or more particles are in the sensing zone simultaneously. If this occurs, they will be read as a single particle. The statistical correction to the number should be no more than a few percent of the total count. To ensure this, the concentration of the dispersion must be very low. The number of particles per unit volume in a dispersion is directly proportional to the solids concentration and inversely proportional to the cube of the particle diameter. Hence, the smaller the size, then the greater is the dilution that has to be made. This can represent a problem with colloidal dispersions of particles made up of soluble and even sparingly soluble materials. The sensitivities of the methods are such that they work best with dispersions containing particles with diameters > 1 μm.

4.1 Electrical Sensing

This was first commercialized with the production of the Coulter Counter which was originally designed simply for counting human red blood cells. It was very quickly developed to enable size distributions to be determined. The principal in this technique is that a dilute dispersion is drawn into a tube through a small cylindrical orifice, with the latter manufactured from sapphire. An electrode is placed on either side of the orifice so that it can be operated as a small conductivity cell. As a particle passes through the orifice, the conductivity changes and as the aperture current is set, the magnitude of the voltage pulse is proportional to the volume of the particle in the sensing zone. The pulse counter is set to progressively lower threshold values during repeated counts so that a cumulative distribution is obtained of the number of particle with a volume greater than the threshold value. The data are then converted to the diameter of the equivalent spherical particle and the numbers adjusted for the coincidence of two particles being within the zonal volume simultaneously and for the background levels of particles in the continuous phase. Filtered media must be used to keep the background down to low levels.

For this conductivity method to work, the final dilution has to be into an electrolyte with a high conductivity. With aqueous systems, a concentration of between 1 and 5 M sodium chloride is required. Non-aqueous systems can be used if a suitable organic electrolyte can be found which is compatible with the dispersion. The very large dilution means that the coagulation rates are slow and so if measurements are done immediately, aggregation does not affect the data. Otherwise, a suitable non-ionic stabilizer for the particles is required. The instrument calculations assume that the particles are non-conducting. If this is not the case, then the data must be corrected to move the distribution to larger-size values.

The range of the instrument is a function of the aperture size. If the particle diameter is < 5 % of that of the orifice, poor discrimination is obtained. If the particle size is > 40 % of the orifice diameter, blockages will occur as when two or three particles arrive simultaneously at the orifice, they can prevent each other from entering. Smaller orifices make this blocking problem more severe but do enable smaller particles to be counted. The usual range of orifice diameters used is from 500 μm down to 30 μm. With the smallest orifices, high aperture currents are needed to produce large enough voltage pulses. The problem then is that of heating of the solution in the aperture and although high electrolyte concentrations can be employed, the noise created by the rapid heating of the fluid in the sensing zone is the limiting factor.

A final point that the operator must bear in mind is that the system is stirred and that the sensing zone is not limited to the actual orifice. This zone extends out a small way from the ends of the orifice so that particles passing through these regions will also be counted. The coincidence correction is based on the random placement of particles in a given volume and a 'cross-flow' through either end of the sensing volume is not allowed for. The stirring then has to be as slow as possible but still sufficient to prevent either sedimentation or creaming.

4.2   Optical Sensing

In this case, we are not referring to equipment that carries out angular light scattering from single particles. Such equipment is available but it should be considered with light scattering in general. Here, we are looking at the change in transmission of light across a defined zone as particles move through that zone. The extinction is proportional to the total amount of light scattered as well as that absorbed.

A rectangular cross-sectional flow chamber is used with a well collimated light source (usually a laser) on one side of the chamber and a photodiode detection system on the other. Apertures can define the optical area very well and so there is not a problem at the edges of the zone as we have with an electrical sensing system. Laminar flow must be used, however, as turbulence must be avoided to prevent local recirculation in the eddies. The counting and pulse height analysis procedures are similar to those used in the electrical zonal system, plus the same problem of a coincidence correction is present. The sensitivity is only in part controlled by the incident intensity as the major factor is the relative cross-sectional area of the particle to the illuminated area. Hence, the particle sizes of the dispersion have to be > 1 μm if the system is to be analysed.

5   SCATTERING METHODS

Colloidal particles scatter electromagnetic radiation and the sizes and shapes of the particles can be calculated by analysing the intensity of the scattered radiation measured as a function of the angle around the scattering volume. The intensity is a function of the particle size and the wavelength of the radiation, as well as the intensity of the latter and the ability of the former to scatter the radiation. The upper end of the wavelength scale is provided by lasers which give plane-polarized light in the wavelength range of 250 to 700 nm. For the lower end of the scales, high-energy X-rays from a synchrotron source and low-energy neutrons from a reactor with a zone cooled with liquid deuterium are employed. The synchrotron gives radiation with an intensity which is about a hundred times brighter than conventional laboratory sources and the wavelength can be varied over a range of approximately 0.05 to 1 nm. The neutrons, tapped from a high-flux reactor, using the cold deuterium moderator to reduce the frequency, have a wavelength range of 0.1 to 1 nm. A mechanical collimator is then used to select a narrow range from this. An alternative source is the use of a circular accelerator with a target which produces a large pulse of neutrons when hit. This is known as a 'spallation' source. Pulses are continuously produced with a range of wavelengths in each pulse. The scattering of a particular wavelength is isolated by using its 'time-of-flight', which is easy to monitor as each pulse provides a start time as a reference. The light scattering photometer is a bench-top piece of laboratory equipment. The X-ray and neutron spectrometers are shared systems in central locations and, of course, much more expensive. They are therefore only occasionally used for size determination alone.

The scattering of light and X-rays occurs from the interaction of the radiation with the electrons of the atoms making up the particles being studied. The neutrons, on the other hand, interact with the nuclei of the atoms. This leads to the extremely important property in that there is a very marked difference in the scattering of neutrons from hydrogen, which just contains a proton as the nucleus, compared to that of deuterium with a neutron and a proton making up the nucleus. Thus, deuterated materials can be introduced into particles, polymers, surfactants or solvents to change the contrast and so hide or bring out structural features of the particles. This is analogous to changing the refractive index in a light scattering experiment and this is very difficult to achieve with simple hydrocarbon-based materials.

We are quite familiar with the interaction of light with matter in that this is used routinely to provide important physico-chemical information. Light can be reflected or refracted by macroscopic objects. It can also be diffracted by the edges of component which are large when compared to the wavelength and we make use of this *Fresnel diffraction* in measuring the sizes of particles at the upper end of the colloid size scale. Light may be absorbed and the energy dissipated as heat or re-emitted at a lower frequency (*fluorescence*).

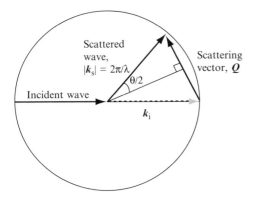

Figure 8.5. Illustration of the geometry employed in a simple scattering experiment (plan view).

However, we will not be dealing with these two phenomena here, with our discussion being centred around the interaction and re-emission of light at the same wavelength. The energy is directly proportional to the frequency of the wave, ω rad s^{-1} (via the Planck constant). There being no dissipation, the scattered wave has the same frequency as the incident wave. Figure 8.5 illustrates the geometry for a simple scattering experiment. Here, the incident wave is scattered through an angle θ, and as there is no energy change, the magnitude of the scattered wave is the same as that of the incident wave, as follows:

$$|k_s| = |k_i| = \frac{2\pi}{\lambda} \tag{8.5}$$

where λ is the wavelength of the radiation. The scattering vector, Q, is given by the following relationship:

$$k_s = k_i + Q \tag{8.6}$$

From the construction shown in Figure 8.5, we can see that the magnitude of the scattering vector is as follows:

$$|Q| = Q = \frac{4\pi}{\lambda}\sin\left(\frac{\theta}{2}\right) \tag{8.7}$$

The scattering vector has units of reciprocal distance and we should note that the distance that we probe in a scattering experiment is about $2\pi/Q$. Measurements at different values of Q may be made by either measuring the scattered radiation at different angles for a fixed wavelength, or by varying the wavelength at a fixed scattering angle. Figure 8.6 shows the scattering angle plotted against the dimension probed by using radiation with wavelengths of 0.35 and 350 nm. The angle used for investigating colloidal particles with neutrons or X-rays is usually $< 10°$. Hence, we refer to

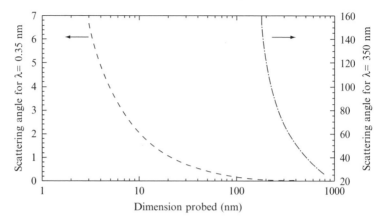

Figure 8.6.   The scattering angle, θ, as a function of the dimension (distance, $2\pi/Q$) for neutrons and light.

small-angle neutron scattering (SANS) and *small-angle X-ray scattering* (SAXS). With light, the range of angles normally used varies from ~ 20 to $160°$, although smaller angles can be probed successfully.

6   ANALYSIS OF SCATTERED RADIATION

We can only provide an outline of this large and important topic here but there are some excellent texts available [1–5] which cover the area in detail. Light scattering from colloidal systems has a long history. The complexity of the analysis, however, increases as the particle radius approaches the wavelength of the radiation, and we will start here with the most straightforward situation.

6.1   Rayleigh Scattering

When the particle radius is < 5 % of the wavelength, a colloidal particle can be adequately treated as a point in the electric field of the light ray. The oscillating electric field causes the outer electrons of the atoms making up the particle to oscillate at the same frequency. We can treat the particle as a single oscillating dipole. Such a dipole radiates light of the same frequency in all directions, where the intensity is a function of the angle to the incident direction and plane of polarization. The first case to consider is that where we are measuring the intensity as a function of the angle in the horizontal plane and the electric field of the light is polarized normal to this plane, i.e. it is vertically polarized. The intensity of the scattered light, $I(\theta)$, relative to the incident intensity, $I(0)$, is given by the following [2]:

$$\left(\frac{I(\theta)}{I(0)}\right)_V = \frac{16\pi^2}{r^2\lambda^4}\left(\frac{\alpha}{4\pi\varepsilon_0}\right)^2 \qquad (8.8)$$

The subscript 'V' indicates that the radiation is vertically polarized and that the scattering angle is measured in the horizontal plane. The intensity is inversely proportional to the square of the distance to the detector, r. The wavelength of the radiation is λ and ε_0 is the permittivity of free space; α is the polarizability of the particle and this is a function of the relative refractive index of the particle, n_2, to that of the suspending medium, n_1, and the volume of the particle, v_p:

$$\alpha_{21} = 3\varepsilon_0\left(\frac{n_{21}^2 - 1}{n_{21}^2 + 2}\right)v_p, \text{ where } n_{21} = \frac{n_2}{n_1} \qquad (8.9)$$

Clearly, as Equation (8.9) indicates, the particle will not scatter light if it has the same refractive index as the medium. Substituting Equation (8.9) into Equation (8.8) and also writing the volume in terms of the particle diameter, D_p, we have:

$$\left(\frac{I(\theta)}{I(0)}\right)_V = \frac{2\pi^4 D_p^6}{r^2 \lambda^4}\left(\frac{n_{21}^2 - 1}{n_{21}^2 + 2}\right)^2 \qquad (8.10)$$

This equation indicates that we will see no angular dependence of the scattered intensity for vertically polarized light.

When the plane of the polarization is horizontal, the angular dependence will change. We will not see the radiation from the particle when we examine the latter along the polarization axis. In this case, the relative intensity is given by:

$$\left(\frac{I(\theta)}{I(0)}\right)_H = \frac{2\pi4 D_p^6}{r^2 \lambda^4}\left(\frac{n_{21}^2 - 1}{n_{21}^2 + 2}\right)^2 \cos^2\theta \qquad (8.11)$$

If the radiation is unpolarized, the incident intensity is equally divided between the vertical and horizontal components and we will see the relative scattered intensity as the sum of the two:

$$\left(\frac{I(\theta)}{I(0)}\right) = \frac{1}{r^2}\left[\frac{\pi^4 D_p^6}{\lambda^4}\left(\frac{n_{21}^2 - 1}{n_{21}^2 + 2}\right)^2 (1 + \cos^2\theta)\right] \qquad (8.12)$$

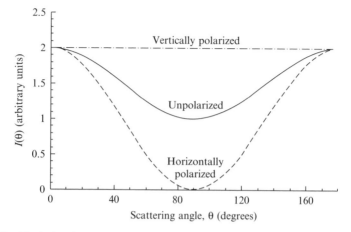

Figure 8.7. Variation in the angular intensity as a function of the scattering angle for a 'Rayleigh scatterer'.

Figure 8.7 shows the variation in angular intensity that we would see for the scattering from a small particle when using unpolarized light, as well as horizontally or vertically polarized light. In Equation (8.12), r is an instrument constant while the term in square brackets is termed the *Rayleigh ratio*, $R(\theta)$. When we have light scattered from a dilute dispersion, each particle contributes and so the intensity as a function of the angle is:

$$I(\theta) = \frac{I(0)}{r^2} R(\theta)\rho \qquad (8.13)$$

where ρ is the number density of particles in the dispersion in the scattering volume. This assumes that the separation between particles is large and so the light scattered by each particle is not subsequently scattered a second or third time by neighbouring particles, that is, there is no multiple scattering. This is easily checked by showing that there is a linear dependence on concentration. Note that Equation (8.12) is very sensitive to the particle size as the particle diameter is given to the sixth power. Hence, if there is a distribution of particle sizes, the larger fractions contribute much more strongly to the scattered intensity than the smaller sizes and the average diameter calculated from the scattering will be weighted to the 'larger end' of the size distribution.

6.2 Rayleigh–Gans–Debye Scattering

As the particle diameter becomes a significant fraction of the wavelength of the radiation, we can no longer treat the particles as 'point-scatterers'. Radiation

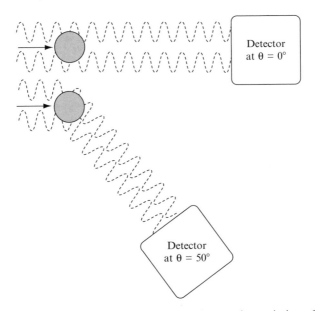

Figure 8.8. Light scattered at 0 and 50°, showing the angular variation of the interference between rays emanating from different parts of a particle.

scattered from different parts of a particle will have different distances to travel to the detector and there will be interference between the waves, the extent of which will be a function of the angle. The problem is outlined in Figure 8.8, in which the particle diameter is $\sim 1.25\lambda$. The intensity of scattered radiation is modified from that calculated with Equation (8.13) by the *form factor*, $P(\theta)$, as follows:

$$I(\theta)_{RGD} = I(\theta)_R P(\theta) \qquad (8.14)$$

where the subscripts 'R' and 'RGD' indicate the intensities from Rayleigh and Rayleigh–Gans–Debye calculations, respectively. The theory is applicable to particles where $(n_2 - n_1)D_p/2\lambda < 1$.

For small deviations from the Rayleigh condition, the form factor can be approximated for any shaped particle [4] by the following:

$$P(\theta) = 1 - \frac{(QR_G)^2}{3} + \dots \qquad (8.15)$$

In the above, R_G is the radius of gyration of the particle and so Equation (8.15) can be used to calculate the angular variation for rods or discs, as well as for spheres. For example, the radius of gyration of a spherical particle with a homogeneous density profile is as follows:

$$R_G = \left(\frac{3}{5}\right)^{0.5} \frac{D_p}{2} = 0.39 D_p \tag{8.16}$$

If we measure the intensity at two angles, we may use Equations (8.14) and (8.16) to determine the particle size. Angles of 45 and 135° are usually chosen for this and the ratio of the two intensities is known as the *dissymmetry ratio*. Now, for our homogeneous sphere we can write the dissymmetry ratio, after substituting for Q from Equation (8.7), as follows:

$$\frac{I(45)}{I(135)} \approx \frac{1 - 1.17(D_p/\lambda)^2}{1 - 6.83(D_p/\lambda)^2} \tag{8.17}$$

Therefore, we can determine the diameter simply from measuring the dissymmetry ratio.

Alternatively, we may use the value of $I(90)$ as a reference intensity (the $\cos\theta$ term in Equation (8.12) becomes zero), and so we can write:

$$I(\theta) \approx I(90)(1 - \cos^2\theta)[1 - (QR_G)^2/3] \tag{8.18}$$

Equation (8.18) gives us the radius of gyration of our particle. For homogeneous particles, the radii of gyration for various shapes are given by the following:

$$\text{spherical ring, radius } a \colon R_G = a$$
$$\text{sphere, radius } a \colon R_G = a\sqrt{3/5}$$
$$\text{thin disc, radius } a \colon R_G = a/\sqrt{2}$$
$$\text{thin rod, length } L \colon R_G = L/\sqrt{12} \tag{8.19}$$

and so we can determine the particle dimension from the ratio of the intensities.

As the particle size increases further from the Rayleigh region, the form factor can no longer be adequately described by using only the first term in the series expansion (Equation (8.15)). For homogeneous spheres of radius a, we may write the form factor as follows [4]:

$$P(\theta) = \{3[\sin(Qa) - Qa\cos(Qa)]/(Qa)^3\}^2 \tag{8.20}$$

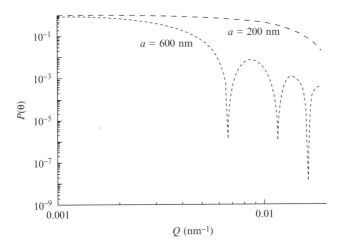

Figure 8.9. Variation of the particle form factor, $P(\theta)$, as a function of Q for two different particle sizes, when using a wavelength of 600 nm.

Figure 8.9 shows the variation of $P(\theta)$ as a function of Q calculated for two different particle sizes when using a wavelength of 600 nm. There is a marked structure in the angular variation of the form factor as the radius approaches the wavelength of the radiation. As the particle size distribution broadens, the sharp troughs become shallower and are smoothed out for a sample with a broad size distribution.

For large particles, i.e. $(n_2 - n_1)a/\lambda \sim 1$, the theoretical analysis, known as the Mie theory [2], is much more complex and is limited to spherical particles. Computer programs are available to match the structure of the angular intensity scan from the theory with that observed experimentally [2, 6]. The problem is not just computational, however, but is also one of measuring the absolute intensities accurately so that a lot of care is required to ensure good data. This is always true for light scattering as small amounts of airborne dust can easily get into our dispersions. As most dust particles are in the upper end of the colloid size domain, the scattering from such contaminants can result in significant errors.

6.3 X-Rays and Neutrons

The above discussion is presented in terms of the refractive indices of the disperse and continuous phases. The analysis is the same for both X-rays and neutrons except that the scattering length density of the particles is used instead of the refractive index. For a particle, the scattering length density is calculated by summing the scattering lengths, b_i, of all of the atoms in the particle. As we noted above, the scattering of X-rays occurs from the electrons of the atom and is therefore the product of the atomic number, z_i, of the atom and the scattering length of an electron (which is 2.8×10^{-15} m). Neutrons, however, are scattered from the nucleus and the scattering length does not vary regularly with the atomic number of the atom. The values are tabulated, of course, and so this does not represent a problem. We can calculate the scattering length densities for neutrons and X-rays of our particles, ρ_{SN} and ρ_{SX}, respectively, as follows:

$$\rho_{SN} = N_m \sum_i b_{N_i}$$
$$\rho_{SX} = N_m \sum_i b_{X_i} = 2.8 \times 10^{-15} N_m \sum_i z_i \tag{8.21}$$

where N_m is the number of molecules per unit volume made up of atoms of type i in the particle, namely:

$$N_m = \frac{\rho_m N_A}{M_m} \tag{8.22}$$

in which ρ_m and M_m are the density and molar mass of the molecules, respectively, and N_A is the Avogadro constant. Similar calculations must be carried out for the continuous phase, of course. Values of the scattering lengths of some selected atoms are given in Table 8.1. We can express the refractive index in terms of the scattering length density and the wavelength as follows [7]:

$$n = 1 - \rho_{SN} \lambda^2 / \pi \tag{8.23}$$

and we would not observe scattering from the particle if its refractive index matches that of the medium. This ability to 'contrast-match' can often be exploited.

Table 8.1. Scattering lengths (10^{-15} m) for neutrons and X-rays of some typical atoms

Atomic species	b_{SN}	b_{SX}
H	−3.74	2.85
D	6.67	2.85
C	6.65	17.1
O	5.81	22.8
Na	3.36	31.3
Si	4.15	39.0
Cl	9.58	48.4
Cd^{2+}	3.7	131.1

By using equations (8.21) and (8.22) with the values taken from Table 8.1, the scattering length density for water is as follows:

$$\frac{10^3 \times 6.022 \times 10^{23}}{0.018}[5.81 - (2 \times 3.74)] \times 10^{-15} = -0.56 \times 10^{14}\,\text{m}^{-2}$$

while that for deuterium oxide is $6.34 \times 10^{14}\,\text{m}^{-2}$. Many hydrocarbon polymers fall in between these values so that polymer particles or coatings can be visualized separately if combinations of hydrogenated and deuterated materials are used. This technique is known as 'contrast-matching'. As an example that may be used to illustrate how we could apply this technique, consider a small polystyrene particle (say, $r \sim 25\,\text{nm}$) coated with a monolayer of dodecanoic acid as a stabilizer. In the experiment to study the coating, we could use d_{23}-decanoic acid. Now, the scattering length density for h_8-polystyrene is $1.41 \times 10^{14}\,\text{m}^{-2}$ and that for the deuterated dodecanoic acid is $5.3 \times 10^{14}\,\text{m}^{-2}$. In a mixture of approximately 25 % D_2O/75 % H_2O, we will have a scattering length density match with the core particle and will 'see' the layer as a hollow spherical shell. On the other hand, when we use a mixture of 85 % D_2O/15 % H_2O we will no longer be able to 'see' the surfactant shell with the neutron beam and will just scatter from the core polystyrene particle. In 100 % water, of course, we will scatter from both the core and the shell. Figure 8.10 illustrates this effect.

It is convenient to express the scattering of neutrons from a dispersion of particles in the following form:

$$I(Q) = A\rho\left(\frac{4\pi a_p^3}{3}\right)^2 (\rho_p - \rho_m)^2 P(Q) \tag{8.24}$$

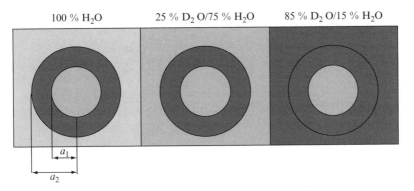

Figure 8.10. Illustration of the 'contrast-matching' of a polystyrene particle with a deuterated dodecanoic acid layer using different H_2O/D_2O mixtures.

where A is the instrument constant which includes the wavelength, the incident intensity and the distance to the detector. (Neutron detectors are made up of two-dimensional arrays of $1 \, cm^2$ measuring elements.) The subscripts ρ_p and ρ_m refer to the scattering length densities of the particle and medium, respectively, a_p is the radius of the particle and ρ is the number density of particles in the dispersion. The equivalent expression for concentric spheres, that is, a dispersion of particles with attached layers, is as follows [8]:

$$\frac{I(Q)}{A\rho(\frac{4\pi}{3})^2} = (\rho_p - \rho_m)\left[3a_1^3\left(\frac{\sin \ Qa_1 - Qa_1\cos \ Qa_1}{(Qa_1)^3}\right)\right]$$
$$+ (\rho_L - \rho_m)\left[3a_2^3\left(\frac{\sin \ Qa_2 - Qa_2\cos \ Qa_2}{(Qa_2)^3}\right) - 3a_1^3\left(\frac{\sin \ Qa_1 - Qa_1\cos \ Qa_1}{(Qa_1)^3}\right)\right]$$

$$(8.25)$$

The first term on the right-hand side of Equation (8.25) represents the scattering from the core particle while the second term is that from the shell with a scattering length density of ρ_L. Clearly, as the scattering length density of the medium is adjusted to equal that of the layer or the particle, we will be able to determine the dimensions of the particle or the layer, respectively.

6.4   The Guinier Approximation

When $Qa \ll 1$, Equation (8.15) provides a reasonable approximation to $P(Q)$, and so we have:

$$I(Q) = A\rho v_p^2(\rho_p - \rho_m)^2\left[1 - \frac{(QR_G)^2}{3} + \cdots\right] \qquad (8.26)$$

Writing this as the exponential form instead of the series:

$$I(Q) = A\rho v_p^2(\rho_p - \rho_m)^2\exp\left[-\frac{(QR_G)^2}{3}\right] \qquad (8.27)$$

which in logarithmic form gives:

$$\ln \ I(Q) = \ln \ I(0) - \frac{(QR_G)^2}{3} \qquad (8.28)$$

Therefore, from a plot of $\ln I(Q)$ versus Q^2 we have the radius of gyration of a particle of arbitrary shape from the initial slope as $-R_G^2/3$, and the intercept as $I(0) = A\rho v_p^2(\rho_p - \rho_m)^2$.

6.5 Porod's Law

This describes the scattering response at high values of Q. As the angles and wavelengths become large, the intensity of scattered radiation varies as Q^{-4}. The limiting equation is as follows:

$$I(Q) \approx 2\pi S(\rho_p - \rho_m)^2 Q^{-4} \tag{8.29}$$

where S is the surface area of the material. It is useful to think in terms of the *fractal dimension* of the surface, d_{sf}. Then the proportionality is:

$$I(Q) \propto Q^{-(6-d_{sf})} \tag{8.30}$$

When we think of a characteristic distance on the surface, r, the area is proportional to r^2, and so the fractal dimension is 2. However, when we have a rough surface there is a three-dimensional character so that the area can be thought of being proportional to r^3, and then $d_{sf} \rightarrow 3$, and therefore:

$$\begin{aligned} I(Q) &\propto Q^{-4}, \text{ smooth surface} \\ I(Q) &\propto Q^{-3}, \text{ rough porous surface} \end{aligned} \tag{8.31}$$

7 NEUTRON REFLECTION

Macroscopic surfaces can be investigated by using a small-angle neutron reflection technique. The intensity of the neutron beam is measured as a function of the incident angle, θ, and of the wavelength. The reflectivity is dependent on the properties of the surface, such as the thickness of any surface layer and its scattering length density. The reflectivity, $R(Q)$, is determined from the variation across the interfacial region, that is, in the z-direction:

$$R(Q) = \frac{16\pi^2}{Q^4} \left| \int \frac{\partial \rho(z)}{\partial z} \exp(iQz) dz \right|^2 \tag{8.32}$$

where we can see the Q^{-4} decay that we expect from a flat surface. The reflectivity is a complex quantity (note the iQz exponential term) as we would expect for any waveform.

This technique has been used effectively to study monolayers of surfactants on water. The water layer can be rendered 'invisible' by using a ratio of D_2O/H_2O of 0.08:0.92 as the scattering length density of the mixture is now zero. The reflection is then just from the monolayer. The scattering length density of the monolayer, ρ_F is:

$$\rho_F = \frac{\sum_i b_{N_i}}{A_m l_m}$$
(8.33)

where A_m is the area per molecule in the layer and l_m is the length of the molecule in the layer (layer thickness). (Note the latter will not necessarily be the length of a stretched surfactant – it is just the thickness of the layer.) We can treat the scattering length density as uniform across our surfactant monolayer so that the reflectivity is now:

$$R(Q) \approx \frac{16\pi^2}{Q^4} \rho_F^2 2[1 - \cos(Qz)]$$
(8.34)

A plot of $R(Q)Q^4$ versus Q varies sinusoidally with a total amplitude of $32\pi^2\rho_F^2$ and the peak at π/l_m. Hence, the length of the surfactant in the monolayer can be determined from the value of Q at the first peak and the area per molecule from the amplitude of this peak, via Equation (8.34).

8   DYNAMIC LIGHT SCATTERING

The time-average scattering that we have discussed so far is elastic scattering in that the frequency of the scattered radiation is the same as the incident value. The particles in the suspension are moving with Brownian motion and we can expect a broadening of the wavelength due to a Doppler shift. In principle, this shift would yield the diffusion coefficient and thus give a measure of the particle size. However, this shift is small and difficult to measure.

The technique that we use to follow the motion is known as *photon correlation spectroscopy* [9], the name of which is indicative of the experimental methodology. Lasers produce coherent light and the equipment is focused to give a small scattering volume. The particles in the dilute dispersion scatter the light and because the light is coherent, the phase relationship is maintained. This means that the dispersion acts like a three-dimensional diffracting array. The structure is random (as the dispersion is dilute and therefore the particles are non-interacting by definition) and we see a random diffraction pattern. As the particles are in motion, the diffraction pattern also moves. This is termed a 'speckle pattern'. From the detector (photomultiplier), we look at the small area defined by the scattering volume and the bright spots or speckles move in and out of our vision at a rate that is dependent on the Brownian motion. The photomultiplier is connected to a photon correlator, and the intensity is measured as a function of time. The total time can be set and the intensity (number of photons) is measured over a series of short time-intervals until the total time is reached. Such a time interval is termed the 'correlation delay time', τ_c. Remember that the intensity is rapidly fluctuating up and down as the speckles move in and out of the field of view. Hence, at two of these times which are close together there is a strong correlation in the intensities but this will not be the case for times which are widely separated. The instrument constructs the correlation function by comparing the product of the number of photons counted initially with those counted in the correlation delay time intervals measured over longer and longer times with the square of the 'long-time' value. The correlation function decays exponentially, as follows:

$$g(\tau_c) = \exp\left(-\tau_c D_s Q^2\right) \tag{8.35}$$

where D_s is the self-diffusion coefficient of the particle. Note that $D_s Q^2$ is just the time taken for a particle to diffuse through a distance of Q^{-1}. The data are plotted as the natural logarithm versus Q^2, which should yield a straight line, and then the diffusion coefficient is found from the slope. The Stokes–Einstein relationship is used to give the hydrodynamic radius of the particle:

$$a_H = \frac{k_B T}{6\pi\eta_0 D_s} \tag{8.36}$$

where η_0 is the viscosity of the continuous phase. Of course, as the dispersion consists of particles of monodisperse hard spheres, the hydrodynamic radius should correspond with that determined by microscopy. If the particles are non-spherical or are covered with a stabilizer layer which has a thickness which is a non-negligible fraction of the radius, the hydrodynamic radius will be larger than that obtained from microscopy measurements.

As the size distribution becomes broader, the range of diffusion times increases in direct proportion. The correlation function then becomes the sum of each of the exponential decaying terms, weighted by the probability of each time occurring. The experimental plot has to be analysed with this in mind as all dispersions have some degree of polydispersity. The problem here is that we cannot obtain a unique solution when we try and invert the summation. (Mathematically, this is termed an 'ill-conditioned' problem.) The method used by each instrument manufacturer is to set up a model which is calibrated against particles of known size. This usually works well for mixtures of monodisperse particles of well separated sizes but may not be so reliable with a single but broad distribution. In addition, as the intensity of scattered light is being measured, the mean will be weighted to the larger sizes.

9 CHARACTERIZATION OF THE ELECTRICAL PROPERTIES OF PARTICLES

In this section, we will concentrate on the properties of the surfaces of colloidal systems rather than bulk properties such as the dielectric behaviour. The surface charge densities of particles can be determined by titration. The larger the particle size, then the more difficult this becomes as the total titratable charge is inversely proportional to the specific surface area. Electrokinetic techniques are used to determine the ζ-potential. The larger part of our discussion will centre around the latter as this is an important factor for determining the interactions between particles.

9.1 Surface Charge By Titration

This technique is appropriate for particles with fixed ionogenic groups on the surfaces of the particles. A good example of such a surface is a polymer latex particle with charge polymer groups. These can be the terminating groups on the ends of the polymer chains or result from a charged copolymer. A good example of the latter are latices which have an considerable fraction of acrylic acid copolymerized in the particles. These latices can be used as modifiers in portland cement formulations.

As it is the strongly attached or covalently bound charge that we are measuring, physically adsorbed material such as weakly adsorbed surfactants have to be removed. In addition, all of the counter-ions to the surface groups must be converted to ones which can be readily titrated, e.g. protons or hydroxyl ions. Free material in solution must also be removed. Treatment with a mixed-bed ion-exchange resin which has its components in the H^+ and the OH^- forms can achieve all of these objectives, leaving acid groups in the hydrogen form, and all salts, acids or bases converted to water.

The dispersion is then titrated with acid or base as appropriate. The 'equivalence-point' is determined either conductometrically or potentiometrically. The experimental difficulty is that dilute acid or bases may have to be used, e.g. at concentrations of 10^{-2} M, and we have to be careful to exclude carbon dioxide. Conductometric data are easier to obtain and give the total charge with the equivalence points being shown by clear changes in the slopes. The latter reflect how strong or weak the acids or bases are. For example, strong acids give a marked negative initial slope as mobile protons are replaced by lower-mobility sodium ions. Weak acids have a slight positive initial slope as the acid groups in the hydrogen form are poorly dissociated and so only a few protons are contributing to the measured conductivity, but all of the sodium ions that have been used to replace them are ionized. Potentiometric data are less easy to analyse but give additional information in terms of the pK_a values of the groups. We should keep in mind with particle surfaces that the ionization of one group is affected by the neighbouring groups, that is, the surface has a polyelectrolyte character and so can have a wide range of pK_a values rather than a single one.

9.2 Electrokinetic Methods

All of these methods rely on the movement of the suspending medium past the charged interface. As such, they can in some cases be applied to macroscopic surfaces in addition to colloidal particles. The property which is determined here is the ζ-potential and not the surface potential. This is because we are making measurements from the relative motion of the electrolyte solution and the surface. The counter-ions that are strongly bound in the Stern layer are assumed to be static on the surface and so the relative motion starts between this region and the hydrated ions and water molecules next to the Stern layer. This gave rise to the concept of a 'shear plane'. This is a convenient simplification for our calculations but we must be aware that it is no better than the modelling assumption that we normally use for a planar particle surface with the charge smeared out uniformly over this. Hence, although the models are satisfactory for distances that are many molecular diameters away from the surface, the precision becomes less at very close distances. Thus, we are left with a potential that we can measure reasonably accurately but at a distance from the surface which will be around 0.5 or 1 nm. The important point though is that it represents a good description of the potential that a test probe would experience as it was brought from a large distance towards the surface. The Stern potential can be much lower and even of reverse sign to the surface potential due to the strong adsorption of multivalent counter-ions or surfactant molecules. There are a number of methods available to determine the ζ-potentials of surfaces. A summary is presented in Table 8.2 and we will briefly discuss these in the following.

Throughout our discussion, we are going to assume that the liquid is Newtonian, that is, the shear stress, σ, is directly proportional to the shear rate, $\dot{\gamma}$, produced in the liquid, and the proportionality constant is the coefficient of viscosity, η. The definitions of these terms are illustrated in Figure 8.11. In general, the fluid velocity gradient is not linear over large distances and it is important to know how this varies with distance as liquid flows in, for example, a capillary tube, which can be used to determine the ζ-potential of macroscopic surfaces. The geometry that we will use for flow through a capillary is shown in Figure 8.12. For steady flow through the tube, the applied force is equal to the viscous drag force:

$$\Delta P \pi r^2 = -\frac{\mathrm{d}v(r)}{\mathrm{d}r}\eta 2\pi r L \tag{8.37}$$

Table 8.2. Examples of electrokinetic techniques used to measure the ζ-potential of particles

Technique	Conditions	Procedure
Streaming potential	Static interface	Move liquid; measure potential
Electro-osmosis	Static interface	Apply potential; measure liquid motion
Sedimentation potential	Moving interface	Particles sediment; measure potential
Electrophoresis	Moving interface	Apply potential; measure particle motion
Primary electroviscous effect	Moving interface and liquid	Measure suspension viscosity
Ultrasonic vibration	Moving interface and liquid	Apply ultrasound; measure AC potential

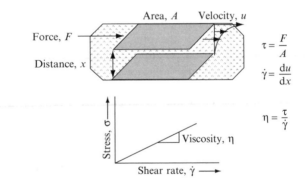

Area, A Velocity, u

Force, F

Distance, x

$$\tau = \frac{F}{A}$$

$$\dot{\gamma} = \frac{du}{dx}$$

Stress, σ

Viscosity, η

$$\eta = \frac{\tau}{\dot{\gamma}}$$

Shear rate, $\dot{\gamma}$

Figure 8.11. Definitions of the shear stress, τ, the shear rate, $\dot{\gamma}$, and the viscosity, η, for a Newtonian liquid. Note that if x is sufficiently small, then the velocity gradient will be linear and there will be a constant shear rate between the surfaces of area A. This is the case for most rotational viscometers.

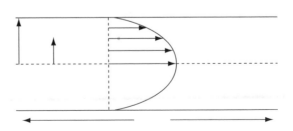

Figure 8.12. Geometry of the flow through a capillary tube.

The boundary conditions for the integration of Equation 8.37 are:

$$v = 0, \text{ at } r = a$$

$$v = v_0, \text{ at } r = 0$$

$$\int_{v(r)}^{0} -\mathrm{d}v(r) = \frac{\Delta P}{2L\eta_r} \int_{r}^{a} r\mathrm{d}r \tag{8.38}$$

This gives us the well-known parabolic velocity profile for a liquid under laminar flow through a tube:

$$v(r) = \frac{\Delta P}{4L\eta}(a^2 - r^2) \tag{8.39}$$

We can use Equation (8.39) to calculate the volume flowing through any element, and by multiplying by the area of that element, $2\pi r\mathrm{d}r$, and then integrating across the radius, we obtain the *Poiseuille equation* for the volume flow rate, Q, through a capillary tube, as follows:

$$Q(r) = v(r)2\pi r\mathrm{d}r \tag{8.40}$$

so that:

$$Q = \frac{\Delta P\pi}{2L\eta} \int_{0}^{a} (ra^2 - r^3)\mathrm{d}r = \frac{\Delta P\pi a^4}{8L\eta} \tag{8.41}$$

9.3 Streaming Potential

The simplest form of this experiment is to use a capillary tube with an electrode compartment at each end, flow electrolyte through the tube and measure the potential developed between the electrodes. This is illustrated in Figure 8.13. A high-impedance voltmeter is required as the resistance between the electrodes is high and care has to be taken to avoid the development of turbulent flow due to too high a flow rate.

In the diffuse layer, there is a higher concentration of counter-ions than co-ions. This, of course, balances the net charge at the Stern plane. The reason a potential develops is that the flow of the liquid moves the ions in the diffuse layer along the tube. There is therefore a net charge flux through the tube and a 'back-current' is produced equal and opposite to this. Far from the wall,

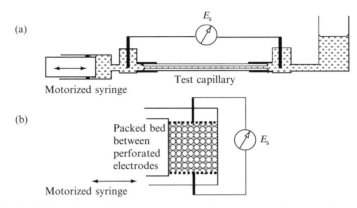

Figure 8.13. Schematics of the apparatus used to measure the streaming potential: (a) test capillary arrangement; (b) cell containing a packed bed of particles or fibres.

the counter- and co-ion fluxes are equal and so only the diffuse layer contributes. Ohm's law is used to calculate the potential developed at the electrodes from this current and the resistance of the contents of the tube. Our problem then is to calculate the number of ions per unit volume at any position in the diffuse layer, the space charge density $\rho_c(r)$, and their velocities. Integration across the tube then gives the total current and hence the *streaming potential*, E_s.

The first step is to write the back-current in terms of the ion flux:

$$i = -2\pi \int_0^a \rho_c(r)v(r)r\,dr \qquad (8.42)$$

Although it is convenient to take the integration out to the tube wall and ignore the thickness of the Stern plane, this, of course, does not imply that we are looking at the surface and not the Stern plane. We can relate the space charge density to the curvature of the diffuse layer potential by using Poisson's equation. In addition, it is acceptable to use the planar form of the equation as long as κa is very large, i.e. we are not considering microscopic pores at low electrolyte concentrations:

$$\frac{d^2\psi(x)}{dx^2} = \frac{-\rho_c(x)}{\varepsilon} \qquad (8.43)$$

where ε is the permittivity of the solution and x is the distance from the charged surface. It is therefore convenient to recast Equation (8.42) in terms of the distance from the wall of the tube, x, using $r = (a - x)$:

$$i = -\frac{\pi\Delta P}{2L\eta}\int_0^a (2ax - x^2)(a - x)\rho_e(x)dx \tag{8.44}$$

Now, as $a \gg x$, we only require the leading term from the expansion of the terms in brackets in this equation and hence we have:

$$i = \frac{\pi a^2 \varepsilon \Delta P}{L\eta}\int_0^a x\frac{d^2\psi}{dx^2}dx \tag{8.45}$$

The boundary conditions for Equation (8.45) are:

$$\psi(x) = \zeta, \text{ at } x = 0$$

$$\frac{d\psi(x)}{dx} = 0 \text{ and } \psi(x) = 0, \text{ at } x = a \tag{8.46}$$

Equation (8.45) is integrated by parts and with the above boundary conditions gives the current as follows:

$$i = \frac{\pi a^2 \varepsilon \Delta P}{L\eta}\zeta \tag{8.47}$$

Noting that the resistance of the tube can be written in terms of molar concentration of the electrolyte, c, and the equivalent conductance, Λ:

$$R = \frac{L}{c\Lambda\pi a^2} \tag{8.48}$$

By using Ohm's law, Equation (8.47) can be written in terms of the streaming potential, E_s, with Equation (8.48). It is also useful to use Equation (8.41) to give the result in terms of the volume flow rate as:

$$E_s = \frac{Q}{c\Lambda}\frac{8L\varepsilon}{\pi a^2}\zeta \tag{8.49}$$

Hence, we can vary Q and plot E_s as a function of Q. This should give a straight line with a slope directly proportional to the ζ-potential. The proportionality constant is $(8L\varepsilon/c\Lambda\pi a^2)$.

We have to be careful about the development of a *standing potential* between the electrodes, indicating slight surface differences and polarization of the electrodes. In order to minimize these effects, the electrodes should be 'shorted' when not being used for measurement and the flow direction should be reversed and the measurement repeated. We will have deviation from a straight line if the flow rate is such that it will result in turbulence, or if a packed bed of fibres or coarse particle is used, the pressure drop is high enough to change the packing and hence the path through the bed.

The capillary (see Figure 8.13(b)) can be constructed of the material in which we are interested or we can coat the internal surface with this material. This may be achieved by physical adsorption of materials such as proteins or surfactants, by chemical grafting of the material, or even by casting the material from solutions or melts. The cell shown in Figure 8.13(b) can contain fibres or coarse particles. If these are fine and low electrolyte concentrations are employed, we should be aware that the conductivity that we might measure would become dominated by the contribution from transport in the diffuse layers, that is, surface conductance, and the calculated ζ-potential would be lower than if this effect is allowed for.

9.4   Electro-osmosis

Electro-osmotic flow can be observed when an electrical potential is applied across a capillary or a porous plug and is due to the motion of ions moving in the potential gradient. Of course, in the bulk there is no net flow as the mass transport due to anions and cations balance out each other. This is not the case when one of the types of charge has a fixed location while the other has not, as is the case with a surface where the diffuse layer ions are free to move freely in a direction parallel to the surface. We will consider the case of a capillary with a charged wall to illustrate the effect. The electrical field strength is the applied potential per unit length of the capillary, E. Consider a small element, dx thick and of unit area. The force on this element due to the applied field on the solvated ions is then:

$$\text{electrical force} = E\rho_c(x)dx \tag{8.50}$$

This force produces the motion of the element and when the flow is steady, it is balanced by the viscous drag force. This drag on the surface of the fluid element is due to the relative velocity change across the element and the stress is determined from the rate of change of the stress across the element multiplied by the distance over which it is changing, as follows:

$$\text{viscous force} = \eta \frac{d^2 v(x)}{dx^2} dx \tag{8.51}$$

Equating the forces for the steady flow condition and substituting for the space charge density by using the planar form of Poisson's equation (Equation (8.43)), and then integrating twice with respect to x to give the fluid velocity a long distance away from the wall, yields:

$$-E\varepsilon \int \frac{d^2\psi(x)}{dx^2} dx = \eta \int \frac{d^2v(x)}{dx^2} dx \qquad (8.52)$$

The boundary conditions for the above equation are as follows:

$$x = 0, \psi = \zeta \text{ and } v(x) = 0$$

$$x = a, \frac{d\psi}{dx} = 0, \psi = 0 \text{ and } \frac{dv(x)}{dx} = 0 \qquad (8.53)$$

and these give the velocity at a distance x away from the wall as:

$$v(x) = \frac{E\varepsilon}{\eta}[\zeta - \psi(x)] \qquad (8.54)$$

When $x \gg 1/\kappa$, the velocity no longer changes with distance and we have 'plug flow' of the liquid through the tube, that is, the velocity profile is flat over most of the tube radius with the change in the region close to the wall. This is a situation which appears similar to 'wall slip' as the bulk of the liquid moves unsheared through the capillary, although it arises from a different origin. The maximum velocity, which we may call the electro-osmotic velocity, v_{eo}, from Equation (8.54), is given by the following:

$$v_{eo} = \frac{E\varepsilon}{\eta}\zeta \qquad (8.55)$$

Two schematic arrangements whereby the electro-osmotic flow in a capillary tube of radius a can be measured are illustrated in Figure 8.14. A packed bed of coarse particles or fibres could be substituted for the capillary in each case. The bed itself would need to be held in place by perforated electrodes. Figure 8.14(a) is the simplest arrangement. Here, the volumetric flow rate, Q, is measured, and this is equal to the electro-osmotic velocity multiplied by the cross-sectional area of the capillary. In addition, Ohm's law can be used to cast the relationship in terms of the measured current and electrolyte concentration:

$$EL = i\frac{L}{c\Lambda\pi a^2} \qquad (8.56)$$

The flow rate is then given by:

$$Q = i\frac{\varepsilon}{c\Lambda\eta}\zeta \qquad (8.57)$$

and a graph of the flow rate versus the current yields the ζ-potential from the slope.

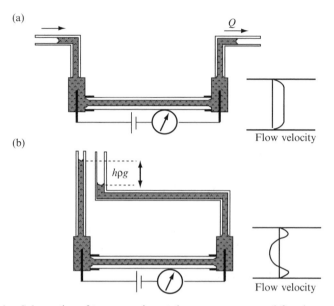

Figure 8.14. Schematics of two experimental arrangements used for the measurement of electro-osmotic flow in a capillary. In the sealed system (b), the liquid is forced to recirculate by moving back along the tube centre.

In Figure 8.14(b), the exit tubes are arranged vertically and the result of the applied field is to produce flow resulting in a difference in height of the liquid in the two limbs. Under steady conditions, the liquid appears static with a difference in the head. This does not mean that the flow near the capillary walls has ceased. What is happening is that the liquid is forced to recirculate back down the centre of the tube. The flow profiles of both arrangements are also shown in Figure 8.14. Making use of the Poiseuille equation (Equation (8.41)), we can equate the two flow rates and the electro-osmotic pressure becomes:

$$h\rho g = i \frac{8L\varepsilon}{\pi a^4 c\Lambda} \zeta \tag{8.58}$$

A plot of the pressure versus the current then enables the ζ-potential to be determined.

As with streaming potential, this technique may be used for materials which can be made in the form of capillaries, and to study the adsorption of surfactants or proteins onto these surfaces, as well as for packed beds of particles and fibres.

9.5 Sedimentation Potential

This is also known by the term *Dorn effect*. As large colloidal particles sediment in an electrolyte, there is relative motion between the particles and the fluid in a similar manner to fluid flowing through a packed bed in a streaming potential experiment. A potential is developed between the top and bottom of the sedimenting region. The velocity of the motion is governed by Stokes law for very dilute systems and is slowed by the 'back-flow' of liquid as the volume fraction of the suspension exceeds 0.1. The experimental difficulties in using this effect to estimate the ζ-potential then are first that the potential is small as the velocity is small, and secondly that in order to vary the velocity, a centrifuge would be required and this makes the measurement of the potential rather difficult. Hence, this is not a methodology that is of more than passing interest.

9.6 Particle Electrophoresis

This is the 'inverse' of the sedimentation potential in that here we apply an electric field and measure the velocity of the particles. This is the most widely used technique for the determination of the ζ-potentials of colloidal particles. It is also the most widely studied of all of the electrokinetic techniques, with a wide variety of cell types being used. Traditionally, cylindrical cells were used with an ultramicroscope arrangement to view the particles. The microscope illuminates the particles at 90° to the viewing direction so that the particles are observed as bright spots due to the light scattered from them. The motion is then measured as a function of the applied field. Currently, laser illumination is employed and the motion is measured by using a variant of photon correlation spectroscopy. The results are then given as the electrophoretic mobility, u_e, which is the velocity per unit field strength:

$$u_e = \frac{v_e L}{E} \text{(in units of m}^2 \text{ s}^{-1} \text{ V}^{-1}\text{)} \qquad (8.59)$$

Large particles will sediment and this contribution can be ignored as long as the electric field and the motion is measured in the horizontal direction. Brownian motion, however, is random and not uni-directional and so cannot be eliminated from the measurement. This effect is superimposed on the measured particle motion and is responsible for a spread of velocities with even the most uniform dispersions. For example, if we have a particle with a diameter of 0.5 μm and $\zeta \sim 70$ mV, we would expect a 5 % spread in the measured electrophoretic mobility values just from Brownian motion alone. If a cylindrical cell is used for the measurements, electro-osmotic flow will occur due to charges on the wall of the cell. If a sealed system is used, measurements can be made at the point of the recirculating electrolyte where there is no net motion of the fluid. There is a cylindrical surface inside the cell where the motion of the liquid in one direction is just counterbalanced by the motion in the other direction and this region is known as the *stationary level*. The net velocity is calculated from summing the electrophoretic velocity and the velocity from the Poiseuille flow in the reverse direction. The null position is at a distance of $a/\sqrt{2}$ from the centre of the tube. The experimental apparatus must be carefully adjusted in order to make measurements at this position. However, measurements can also be made between electrodes in an optical cuvette where there is no cell surface between the electrodes. This completely avoids the problems stemming from the use of capillary cells and cells can be placed in dynamic light scattering equipment with very little modification.

☐ The Mobility at κa > 100

The diffuse layer is small compared to the radius of the particles. This enables us to treat the diffuse layer as planar relative to the particle surface and both the lines of force from the electric field and the fluid flow lines are then taken as being parallel to the particle surface. So, we have the condition for a large particle at a moderate-to-high electrolyte level. This is the case which is nearly always assumed in the software that is supplied with most pieces of equipment when ζ-potential is being evaluated as opposed to mobility.

Figure 8.15 illustrates the model which is employed. The particle is moving at a velocity v_e through the electrolyte solution due to the application of an electric field and hence there is a velocity gradient close to the particle surface and we have an analogous situation to that which we have discussed for electro-osmosis. When we consider a small element of unit area at a distance

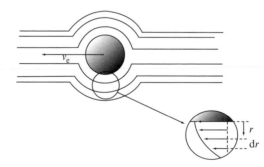

Figure 8.15. The model for electrophoretic mobility used to derive the Smoluchowski equation (Equation (8.64) below) with simplified flow lines around a particle moving with a velocity v_e due to an applied field. The particle is assumed to be non-conducting and the electrical field is also distorted in order to be parallel to the surface of the sphere.

r from the surface, the change in the viscous drag on this element is the result of the change in the velocity gradient across it. So, by integrating with respect to *r* we get the viscous stress at the surface of the particle:

$$-\eta \int \frac{d\dot{\gamma}}{dr} dr = -\eta \int \frac{d^2 v(r)}{dr^2} dr \tag{8.60}$$

The negative sign is just indicating that the velocity decreases as we go further away from the particle surface. Now, the electrical force is the charge multiplied by the field strength, while the total charge on the particle, σ_T, is the charge density at the shear plane multiplied by the area of the particle:

$$E\sigma_T = -E4\pi a^2 \varepsilon \frac{d\psi(r)}{dr}\bigg|_{r=0} \tag{8.61}$$

The planar form of Poisson's equation (Equation (8.43)) can be used as the diffuse layer dimension is so much smaller than the radius. When there is steady motion, we equate the forces. To do this, we must multiply the viscous stress by the surface area of the particle:

$$E4\pi a^2 \varepsilon \frac{d\psi(r)}{dr}\bigg|_{r=0} = 4\pi a^2 \eta \frac{dv(r)}{dr} \tag{8.62}$$

In order to calculate the velocity, we must integrate with respect to *r*, giving:

$$E\varepsilon\zeta = \eta v_e \tag{8.63}$$

by using the boundary conditions, $r = 0, \psi(r) = \zeta$ and $v = v_e$. In terms of the mobility, we have:

$$u_e = \frac{\varepsilon\zeta}{\eta} \tag{8.64}$$

This is the *Smoluchowski equation* for the electrophoretic mobility of colloidal particles. Note that there is no size term in the equation as the surface area terms cancel out.

❏ THE MOBILITY AT κ ≪ 1

In this limit, the particle is assumed to act as a point charge so that the lines of force of the electric field are unperturbed by the particle. Again, as we have steady motion we balance the forces on the particle. In this model, the drag on the particle is the sum of two terms. The first of these is taken as the velocity multiplied by the Stokes drag factor, i.e. $v_e 6\pi\eta a$. The second force slowing the particle motion is the electrical force on the ions of the diffuse layer as these will try and migrate in the opposite direction to the particle as their charge is of opposite sign. So, in balancing the forces we can write:

$$\sigma_T E = 6\pi\eta a v_e + Ef(\rho(r)) \tag{8.65}$$

Note that r is the distance from the particle centre. The electrophoretic velocity is now:

$$v_e = -\frac{E4\pi a^2 \varepsilon}{6\pi\eta a} \left.\frac{\mathrm{d}\psi(r)}{\mathrm{d}r}\right|_{r=a} - \text{(a correction term)} \tag{8.66}$$

The correction term is due to the motion created by the movement of the diffuse layer ions resulting from the applied electric field. It is straightforward to calculate the force due to the field on a thin spherical shell in the diffuse layer as:

$$4\pi r^2 E\rho(r)\mathrm{d}r = 4\pi r^2 E\varepsilon \frac{\mathrm{d}^2\psi(r)}{\mathrm{d}r^2} \tag{8.67}$$

This force would induce a motion so that there would be a relative velocity, $\mathrm{d}v(r)$, between layers which would produce the force between adjacent layers, and so approximating each layer to a continuous shell and balancing the drag, the result is:

$$6\pi\eta r \mathrm{d}v(r) = 4\pi r^2 E\varepsilon \frac{\mathrm{d}^2\psi(r)}{\mathrm{d}r^2} \tag{8.68}$$

Integrating with the boundary condition that at large distances the potential gradient is zero and rearranging to give the liquid velocity at a long distance from the particle, v_1, yields:

$$v_1 = \frac{2E\varepsilon}{3\eta} \int \frac{\mathrm{d}^2\psi(r)}{\mathrm{d}r^2}\mathrm{d}r \tag{8.69}$$

We can now substitute Equation (8.69) into Equation (8.66) and with integration by parts with the boundary condition that at $r = a$, $\psi(r) = \zeta$:

$$v_e = \frac{2\varepsilon E\zeta}{3\eta} \tag{8.70}$$

and expressing this in terms of the electrophoretic mobility gives:

$$u_e = \frac{2\varepsilon\zeta}{3\eta} \tag{8.71}$$

This limiting equation is the *Hückel equation*. The mobility that it yields is less than that we would expect from the Smoluchowski equation. However, both are limiting approximations and we require a better estimate for many colloidal dispersion when we have $1 < \kappa a < 100$. Over most of the size range, we cannot treat the particle as a point charge in a uniform field but we have to allow for the distortion of the electric field around the particle. This means that the field is variable with position around the particle and only when the particle is very large can the field be considered to have lines of force parallel to the surface at all points (as in the Smoluchowski model). We need a distortion which is a function of the radius. The local field acts on the diffuse layer ions and alters the retardation correction to the particle velocity [1, 10].

$$u_e = \frac{2\zeta\varepsilon}{3\eta}[1 + Kf(\kappa a)] \tag{8.72}$$

where K is calculated from the conductivity of the particles and the medium, and for non-conducting particles, $K = 1/2$. This is the *Henry equation*. The function of κa can be written as a power series. For example, for $\kappa a > 5$:

$$\frac{f(\kappa a)}{2} = \frac{1}{2} - \frac{3}{\kappa a} + \frac{25}{(\kappa a)^2} - \frac{220}{(\kappa a)^3} + \cdots \tag{8.73}$$

This function provides a smooth transition from the Hückel result to the Smoluchowski result as κa increases to values of ~ 100.

However, there also needs to be a correction for the distortion of the diffuse layer from spherical symmetry. A full analysis requires numerical computation as there is no simple analytical result. Some of the results of such calculations from the literature were published in tabular form by Ottewill and Shaw [11] and these show that for a 1:1 electrolyte the Henry equation (Equation (8.72)) provides an adequate description, although this is not the case with multivalent species such as 2:2 electrolytes, for example, where there can be considerable deviations in the intermediate κa regime and then a numerical computation is required [12].

9.7 Electroacoustics

We have already seen how a DC potential produces the motion of an electrolyte past a surface. Now, if an AC potential is applied to a dispersion we can expect the particles to follow the oscillating field as will the diffuse layer distortion. When the frequency is increased to levels in excess of $\sim 1\,\mathrm{MHz}$, the particle inertia can become large enough to prevent significant particle motion although the ions can still respond and a pressure wave is produced which can be monitored with an ultrasonic transducer. Alternatively, an ultrasonic vibration can be applied and the relative motion of the diffuse layer produces an AC potential (cf. the streaming potential). When the frequencies exceed $\sim 20\,\mathrm{MHz}$, the diffuse layer can no longer respond, the signal is lost and the particle appears to be uncharged. The process has been analysed in detail [13] and is used to determine a dynamic mobility, u_{d}. In addition, the frequency variation may be used to determine the mean particle diameter.

The current commercially available equipment applies an ultrasonic vibration and measures the resultant AC potential. The particle inertia is a function of the particle size but also the difference in the density, $\Delta\rho$, between the particle and the medium. This technique operates most efficiently with dense particles. In addition, the amplitude of the signal is proportional to the volume fraction of the dispersion as the higher the volume fraction, then the more oscillating dipoles there are and so the stronger the signal. We may use this to advantage as there are no optical requirements as we had in electrophoretic measurements, and work at moderate volume fractions can be carried out when the dispersions are opaque. The dynamic mobility is given by:

$$u_{\mathrm{d}} = \frac{\varepsilon\zeta}{\eta} G(\varphi, \Delta\rho) \tag{8.74}$$

where G is a correction factor which is a function of the dispersion concentration, as well as the particle and medium densities, and φ is the volume fraction. The current state of the theory assumes that the diffuse layers around each particle do not interact with those of the neighbouring particles. This means, however, that we are limited to dilute dispersions and we must check that the mobility is linear with concentration. For large particles and thin diffuse layers, we could still work with a high-solids content. However, with small particles and dilute electrolyte concentrations, we can often be limited to $\varphi < 0.1$.

10   VISCOSITIES OF DISPERSIONS

Measurement of the viscosity of a dispersion is a useful method of characterizing the particles in that dispersion. The viscosity is, of course, sensitive to the concentration of particles but in addition to this it is a function of particle charge and shape, as well as the dimensions of any adsorbed layers. Capillary viscometry can be precise and is most suitable for use with dilute dispersions. This is because, as we have noted above, the fluid velocity has a parabolic profile so when we calculate the shear rate from $dv(r)/dr$ we find that there is a linear change of shear rate from zero at the tube centre to a maximum value at the wall. For example, a wall shear rate in excess of $10^3\,s^{-1}$ is common. As a result, we should avoid systems which are non-Newtonian; that is, there is not a linear dependence of shear stress on the shear rate. However, this is not normally a problem with dilute dispersions.

10.1 Dependence on Volume Fraction

Particles in a flowing fluid produce a dilation of the field as the fluid flows around them. Figure 8.16 illustrates schematically the flow around single particles and also how pairs of colliding particles interact. Particles move at the velocity of the streamline in line with the particle centre. The fluid flowing past the upper surfaces of the particles in this figure is moving more rapidly than the particles and conversely that at the lower surfaces is moving more slowly. This produces a rotation of the particle and a rotation rate equal to half of the shear rate, $\dot{\gamma}/2$. This is known as the *vorticity* of the shear field. Figure 8.16 also shows the manner in which colliding pairs of particles interact with the flow field. The 'near-field' interaction is how we first think of a collision in that the particles come together and rotate as a single unit. The 'far-field' interaction though must also be included. In this case, the particle centres are not on a collision trajectory but are close enough so that the dilation of the flow around one particle is disturbed by that around the other. These interactions with the shear field enhance the energy dissipation rate and we measure this as an increase in the viscosity. The hydrodynamics of the interactions are well established and we may write the viscosity of dispersions of hard spheres in shear flow as follows:

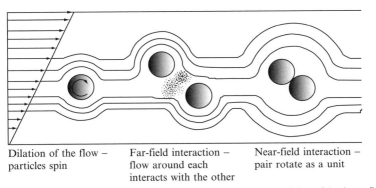

Dilation of the flow –   Far-field interaction –   Near-field interaction –
particles spin   flow around each   pair rotate as a unit
                         interacts with the other

Figure 8.16. Schematic illustration of the interactions of particles with shear flow.

$$\eta(0) = \eta_0[1 + 2.5\varphi + 6.2\varphi^2 + O(\varphi^3)] \qquad (8.75)$$

where φ is the volume fraction of the dispersion and η_0 is the viscosity of the continuous phase. The viscosity of the dispersion is $\eta(0)$ and we are using the '0' to indicate the low shear limiting behaviour so that the spatial arrangement of the particles is not perturbed by the shear rate. This condition means that the diffusional time for a particle, τ, is less than the characteristic shear time, so that $\tau\dot{\gamma} < 1$. The first coefficient in Equation (8.75) was derived by Einstein [14] and the second by Batchelor [15]. We do not have a rigorous hydrodynamic derivation of the coefficient for the three-body collision and the equation is therefore limited to dispersions where $\varphi < 0.1$.

Many flows have an extensional component and some can be mainly extensional with only a small shear component. When a fluid is sprayed, there is a large extensional contribution as the fluid is forced through a small nozzle. This is also the case in blade and roller coating. For extensional flow, the appropriate relationship is as follows [15]:

$$\eta(0) = \eta_0[1 + 2.5\varphi + 7.6\varphi^2 + O(\varphi^3)] \qquad (8.76)$$

This indicates that we can expect differences in the viscous behaviour of dispersions of moderate-to-high concentrations in different types of flow as then the two- and higher-body interactions dominate the response. It is also important to note that it is never possible to achieve a steady extensional flow and we must be aware that the residence time in the extension regime should also be considered. A review of dispersion viscosity can be found in the texts by Hunter [1] and Goodwin and Hughes [16].

Every deviation from the particles being hard spheres results in a change to the Einstein and Batchelor coefficients. Hence, the dispersion viscosity may be used to give information on adsorbed layers, particle charge, particle shape and the fluid nature of the particles. The experimental problem is that we must work with dilute dispersions in order to have rigorous hydrodynamic descriptions and this means that the variation in the viscosity can be quite small. We can re-write Equation (8.75) in the general form:

$$\frac{\eta(0)}{\eta_0} = 1 + [\eta]\varphi + k_H[\eta]^2\varphi^2 + \cdots \qquad (8.77)$$

where $[\eta]$ is the intrinsic viscosity and k_H is the *Huggin's coefficient* [16]. It is usual to express this in linear form, as follows:

$$\eta_{red} = \left[\frac{(\eta(0)/\eta_0) - 1}{\varphi}\right] = [\eta] + k_H[\eta]^2\varphi \qquad (8.78)$$

A plot of the reduced viscosity, η_{red}, versus the concentration then yields the intrinsic viscosity and deviations from '2.5' indicate deviations from hard-sphere behaviour. However, even if that difference is 50%, this will only change the viscosity of a dispersion at a volume fraction of 0.05 by $\sim 6\%$. Now, if we are using a capillary viscometer when we are measuring an efflux time of $\sim 200\,s$, then 6% would mean a change of 12 s. We would obviously like to achieve good precision so that would mean that subsequent times should agree within $< 0.5\,s$ at worst. This is a very demanding experiment.

10.2 Adsorbed Layers

Stabilizer layers are often present on colloidal particles. These may be short-chain surfactants or polymers. In order to optimize the stabilizing properties, the surface concentrations are maximized and hence the layers cannot be regarded as 'free-draining'. This means that the fluid must flow around the layer and not through it. With polymer layers, this will be an approximation as there can be some fluid motion within the layer, although much restricted. At the outer periphery, where the population is the longer tails, significant flow will occur. Thus, the layer thickness must be termed a 'hydrodynamic' thickness and this may be an underestimate of the particle–particle interaction thickness. The way the adsorbed layer is included is by using an *effective volume fraction*, φ':

$$\varphi' = \varphi \left(\frac{a + \delta}{a} \right)^3 \tag{8.79}$$

in Equation (8.75) where δ is the adsorbed layer thickness. In terms of the intrinsic viscosity, we could write:

$$[\eta] = 2.5 \left(\frac{a + \delta}{a} \right)^3 \tag{8.80}$$

An adsorbed layer thickness of 10% of the particle radius will increase the intrinsic viscosity by 33%. As most stabilizer layers have $\delta < 10\,nm$, this method for determining the thickness is limited to particles with a radius less that $\sim 150\,nm$ if the data obtained are going to have reasonable precision.

10.3 Fluid Droplets

When the disperse phase is a fluid, the forces exerted by the shear field can deform the particle. The force maxima are at 45° to the flow direction and so the droplets becomes elongated as the shear force varies from a compression to a tension as we scan the angle to the shear direction from $0°$ through $45°$ to $135°$. As well as the continuous deformation of the droplet, the field vorticity results in the outer surface moving continuously around the droplet, resulting in some circulation within. Taylor [17] carried out the analysis for a droplet where the interface allows the ready transfer of energy across, that is, it is thin and not rigid. In this case, the circulation of the fluid inside the drop resulted in a decrease in the rate of energy dissipation compared to that for rigid particles, and with the viscosity of the internal phase being η_i we have the intrinsic viscosity as:

$$[\eta] = 2.5\left(\frac{\eta_i + 0.4\eta_0}{\eta_i + \eta_0}\right) \tag{8.81}$$

For the above equation, we can consider the following limiting conditions: $\eta_i = \infty$, $[\eta] = 2.5$; $\eta_i = \eta_0$, $[\eta] = 1.75$; $\eta_i = 0$, $[\eta] = 1$.

In many emulsions, the viscosities of the two phases are similar and so the middle condition will be reflected. The last condition is relevant to gas bubbles and is what we would find from measuring the viscosity of low-quality foams, i.e. dilute foams. In the case of a particle with a thick, low-density, polymer layer, we can expect some analogous behaviour in that some distortion of the layer could occur and some motion of the liquid within the layer would decrease the overall rate of energy dissipation when compared to a rigid particle the same size as the particle plus polymer layer, and the intrinsic viscosity would be less than 2.5. Interpreting the viscosity then to give the layer thickness would clearly lead to an underestimate. However, the detailed hydrodynamics would be rather different and so the analogy should not be taken as far as attempting to estimate a 'layer viscosity'.

10.4 Electroviscous Effects

There are a number of electroviscous effects and these have been reviewed most recently by van de Ven [18]. The first one identified, and hence known as the *primary electroviscous effect*, was the effect of the distortion of the diffuse layer caused by the flow of liquid past the particle. The similarity to electrophoresis and electroacoustics immediately comes to mind. The result for the intrinsic viscosity for a large κa and small potentials is:

$$[\eta] = 2.5 + \frac{6(\varepsilon\zeta)^2}{K\eta_0 a^2} \tag{8.82}$$

where K is the specific conductance of the electrolyte solution. This equation was derived by von Smoluchowski and has the same restrictions as his equation for the electrophoretic mobility (Equation (8.64)). As the value of κa falls to below 100, we should have a variation with κa and there are analyses available which have analytical approximations, although the solution for any potential and any κa value must be carried out by numerical calculation [19]. However, under most conditions the increase in the value of the intrinsic viscosity above the value of 2.5 is small and require great precision in the viscosity measurement if an estimate of the ζ-potential is required. The exception is for small particles in dilute electrolyte and then a numerical calculation is required.

The interaction between two charged particles in flow leads to a change in the second coefficient in Equation (8.75). This effect is known as the *secondary electroviscous effect*. The most complete treatment was given by Russel *et al.* [20], which gives us:

$$\frac{\eta(\dot{\gamma})}{\eta_0} = 1 + [\eta] + \left[2.5 + \frac{3}{40}\left(\frac{r_0}{a}\right)^5\right]\varphi^2 \qquad (8.83)$$

where r_0 is the centre-to-centre distance between the particles as they collide. This parameter is calculated from the balance between shear forces pushing the particles together and the electrostatic repulsion from the particle surfaces. Thus, r_0 will vary from $\sim 2a$ at very high shear to much larger values at low shear. Of course, at very low shear, Brownian forces are still operating and they will control the closest distance of approach. In many stable colloidal systems, we find that at a surface separation of $\sim 5/\kappa$ there is $\sim 1k_BT$ of repulsive energy and so as a first estimate we should have $r_0 \sim 2a + 5/\kappa$ as a low-shear limit. Note that we now have a shear dependence, that is, the dispersion is non-Newtonian and this means that we will no longer have a simple interpretation of the flow times as the dispersion viscosity due to the large variation of shear rate across the capillary.

Polyelectrolytes are effective stabilizers of colloidal dispersions as they provide an 'electrosteric' barrier to aggregation. The charged groups along the chains of polyelectrolytes repel each other and cause the chains to take up an expanded conformation. Added electrolyte and, if the charged groups are weak acid groups, changes in pH cause changes in the conformation. This is known as the *tertiary electroviscous effect* as the dimensions of the polymer affect the value of the viscosity. When the polymer is attached to a particle surface, the adsorbed layer can be expanded or contracted by altering the chemical environment – thus δ in Equation (8.80) will change.

10.5 Particle Shape

The rotation of anisometric particles in a shear field is periodic and the orbit depends on the initial orientation. The hydrodynamic analysis is available for prolate and oblate ellipsoids [18] but not for particles of arbitrary shape. In order to introduce the problem, we shall consider the limiting rotational orbits of a rod in a shear field. When the major axis of the rod is aligned with the flow direction, the rod will rotate 'end-over-end' with a constantly changing angular velocity dependent on the angle of the axis to the shear plane. The rod effectively 'flips' over. This orientation is one which corresponds to a high energy dissipation rate compared to a sphere of the same volume. However, the rod could be aligned with its major axis in the shear plane but perpendicular to the flow direction. In this orientation, it simply 'rolls' and this corresponds to a lower rate of energy dissipation than for a similar volume sphere. Rotary Brownian motion will tend to randomize the orientations and so we have to average over all possible configurations. If the Brownian motion is weak compared to the shear field, there are solutions available in the literature [21]. The calculations indicate that axial ratios > 5 are required before the increase in the intrinsic viscosity becomes large enough to be measured with precision. However, when the Brownian motion is strong compared to that from the shear field, the effect is stronger as the time aligned with the flow is reduced. In addition, we find that *shear-thinning* occurs when the rotational rates are of the same order. Thus, for an axial ratio of 5, the intrinsic viscosity increases by 30 % at high shear rates while the increase is 100 % at low shear rates [21].

REFERENCES

1. R. J. Hunter, *Foundations of Colloid Science*, Vol. I, Oxford University Press, Oxford, UK, 1987.
2. M. Kerker, *The Scattering of Light and other Electromagnetic Radiation*, Academic Press, New York, 1969.
3. H. C. van de Hulst, *Light Scattering by Small Particles*, Wiley-Interscience, New York, 1957.
4. A. Guinier and G. Fournet, *Small Angle Scattering of X-Rays*, Wiley-Interscience, New York, 1955.
5. P. Lindner and Th. Zemb (Eds), *Neutron, X-Ray and Light Scattering*, North Holland, Amsterdam, 1991.
6. R. L. Rowell, R. S. Farinato, J. W. Parsons, J. R. Ford, K. H. Langley, J. R. Stone, T. R. Marshall, C. S. Parmenter, M. Seaver and E. B. Bradford, *J. Colloid Interface Sci.*, **69**, 590 (1979).
7. R. J. Hunter, *Foundations of Colloid Science*, Vol. II, Oxford University Press, Oxford, UK, 1989.
8. R. H. Ottewill, 'Small Angle Newton Scattering', in *Colloidal Dispersions*, J. W. Goodwin (Ed.), The Royal Society of Chemistry, London, 1982, pp. 143–164.
9. P. N. Pusey, 'Light Scattering', in *Colloidal Dispersions*, J. W. Goodwin (Ed.), The Royal Society of Chemistry, London, 1982, pp. 129–142.

10. R. J. Hunter, *The Zeta-Potential in Colloid Science*, Academic Press, London, 1981.
11. R. H. Ottewill and J. N. Shaw, *J. Electroanal. Chem.*, **37**, 133 (1972).
12. R. W. O'Brien and L. R. White, *J. Chem. Soc., Faraday Trans. 2*, **74**, 1607 (1978).
13. R. W. O'Brien, *J. Fluid Mech.*, **190**, 71 (1988).
14. A. Einstein, *Ann. Physik.*, **34**, 591 (1911).
15. G. K. Batchelor, *J. Fluid Mech.*, **83**, 97 (1977).
16. J. W. Goodwin and R. H. Hughes, *Rheology for Chemists – An Introduction*, The Royal Society of Chemistry, Cambridge, UK, 2000.
17. G. I. Taylor, *Proc. R. Soc. London, A*, **146**, 501 (1934).
18. T. van de Ven, *Colloidal Hydrodynamics*, Academic Press, New York, 1989.
19. I. G. Watterson and L. R. White, *J. Chem. Soc., Faraday Trans. 2*, **77**, 1115 (1981).
20. W. B. Russel, D. A. Saville and W. R. Schowalter, *Colloidal Dispersions*, Cambridge University Press, Cambridge, UK, 1989.
21. L. G. Leal and E. J. Hinch, *J. Fluid Mech.*, 46, 685 (1971).

Chapter 9

Concentrated Dispersions

1 INTRODUCTION

Much of this volume has dealt with dilute colloidal dispersions where we have considered the properties of single particles or two particles interacting with each other. However, as we increase the concentration of a dispersion, multi-body interactions become increasingly important and we have a condensed phase. This occurs when the interparticle forces produce a structure which is space-filling. The forces may be strongly attractive as occurs with clays used for the manufacture of ceramics, but structures may be due to weakly attractive forces or indeed simply to interparticle repulsion. We saw early in this volume how concentrated aggregates of surfactant molecules produce three-dimensional structures and these are just examples of colloidal condensed phases.

The macroscopic properties of the structures formed by concentrated colloidal dispersions takes us into the behaviour of thickened liquids and gels. Such systems are becoming increasingly referred to as *complex fluids, soft matter* or *soft solids*. There are two important questions that have to be addressed with such materials. The first of these is 'what is the structure?'. Intimately connected with this is how it may be determined. The second question is 'what are the physical properties?'. The latter question is primarily the one of rheological or handling properties of the dispersion. The first question can be successfully addressed for model colloidal dispersion such as monodisperse spheres, while in answering the second, we can always make measurements to record the behaviour which, although often of critical importance to commercially important systems, may be difficult to predict for any but the more simple systems. However, the work on model systems has greatly increased our understanding and is an important guide to the interpretation of more complicated systems.

Colloids and Interfaces with Surfactants and Polymers – An Introduction   J. W. Goodwin
© 2004 John Wiley & Sons, Ltd   ISBN: 0-470-84142-7 (HB)   ISBN: 0-470-84143-5 (PB)

2 THE STRUCTURE OF CONCENTRATED DISPERSIONS

The scattering of radiation as a means of providing information on particle size etc. was described in Chapter 8. The scattering of X-rays by organized molecular structures as opposed to individual molecules is a concept that is generally familiar. This is also the case for the scattering of X-rays, neutrons and light from concentrated colloidal systems. The angular intensity of scattered radiation is the product of that scattered from the individual particles and that from the periodic structure in the dispersion. So, for a dispersion of monodisperse spheres we may write:

$$I(Q) = (\rho_p - \rho_m)^2 N_p V_p^2 P(Q) S(Q) \qquad (9.1)$$

To obtain $S(Q)$ from the intensity data, we measure the intensity of a dilute dispersion where $S(Q) = 1$, and then divide the $I(Q)$ which we obtained for the concentrated system with the values for the dilute one scaled by the ratio of the concentrations. The value of $S(Q)$ varies with Q in a periodic manner when Q^{-1} is of a similar magnitude to the periodicity in the structure of the dispersion. We are used to observing Bragg peaks for the scattering from a highly ordered structure such as we find with a crystal. We have an analogous situation with concentrated dispersions although the interparticle forces often allow much more motion and the peaks tend to be broader and may be similar to the peaks that we find from molecular liquids. The structure is dynamic (from the Brownian motion) and we are observing an average over all of the structures.

The problem then is to go from the structure factor to a description of the spatial arrangement of the particles. In principle, we should be able to carry out a Fourier-transform of the measured structure factor to calculate the pair distribution function which gives us the local density of particles with reference to a central particle, that is, it is simply the probability of finding a particle at any distance r from the centre of a reference particle. This inversion may not be straightforward [1] and is not the usual route. It is more common to either calculate a pair distribution function and then use that to determine an $S(Q)$ for comparison or to use a computer simulation. Figure 9.1 shows structure factors calculated [1, 2] for 'hard-sphere' fluids. This figure serves to illustrate some general points. The first is that as systems become more concentrated, the periodic structure becomes increasingly well-defined. The second point is related to the behaviour at low Q values. Note from this figure how $S(0)$ is greater for the lower-concentration system. $S(0)$ is equal to the osmotic compressibility of the system [1, 3] and it should be no surprise that a more concentrated system is more difficult to compress – in other words, it is less *compliant*. The relationship between the pair distribution function $g(r)$ and the structure factor is as follows [1, 3]:

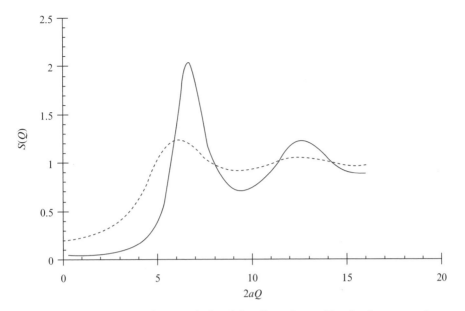

Figure 9.1. The structure factor calculated for dispersions of hard spheres at volume fractions of 0.2 (——) and 0.4 (----). Note that as $Q \rightarrow 0$, $S(Q)$ reduces at higher concentrations, indicating that the compressibility is lower, i.e. the modulus is higher.

$$S(Q) = 1 + \frac{4\pi\rho_0}{Q} \int_0^\infty r[g(r) - 1] \sin{(Qr)} dr \qquad (9.2a)$$

where $\rho_0(= N_p/V)$ is the average number density of particles in the dispersion. Now, the Fourier-transform of this equation gives the pair distribution function, $g(r)$ [1, 3]:

$$g(r) = 1 + \frac{1}{2\pi^2 r \rho_0} \int_0^\infty [S(Q) - 1]Q \sin{(Qr)} dQ \qquad (9.2b)$$

where the limits of $g(r)$ are: $r \rightarrow 0$, $g(r) \rightarrow 0$; $r \rightarrow \infty$, $g(r) \rightarrow 1$.

The shape of the pair distribution function is dependent on the details of the interactions between the particles and is related to the potential of mean force [1]:

$$g(r) = \exp\left(-\frac{\Phi(r)}{k_B T}\right) \qquad (9.3)$$

The potential of mean force is the reversible work done in bringing two particles together from an infinite separation. Of course, in the dilute limit this is just the colloid pair potential but in a concentrated system there are all of the interactions with other particles along the way. Hence, the potential of mean force has an oscillatory character and this shows in the structure of g(r).

The structures illustrated in Figure 9.2 are for two common concentrated colloidal situations. The structure for the weakly attractive system (Figure 9.2(a)) shows the particles in close contact but although the packing is dense it is not highly ordered. This would be typical of say a depletion-flocculated system or a sterically stabilized system where there is no long-range repulsion. The pair distribution function, g(r), is also shown and the shape is typical for a system which is weakly attractive with a very sharp peak corresponding to the first nearest-neighbour shell at $r/2a = 1$. Figure 9.2(b) presents the situation for a system with long-range repulsion. The separation between the particles would correspond to a volume concentration of $\sim 25\%$. Here, the first nearest-neighbour peak corresponds to $\sim r/2a = 1.2$. There is slightly more short-range order here, but with such a large surface–surface separation (about $0.4a$) the Brownian motion results in a rapid fall in the structure relative to the central particle. As the concentration is increased with this sort of system, we can have colloidal crystals formed. However, the usual structures are nearer to a 'colloidal glass'. This is because a considerable time is required for the reorganization to occur to produce well-defined crystals

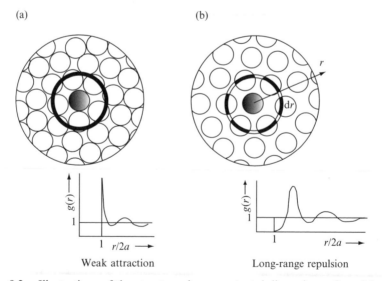

Figure 9.2. Illustrations of the structures in concentrated dispersions of particles with (a) weak attraction, and (b) long-range repulsion between the particles. The corresponding pair distribution functions are also shown.

(usually face-centred-cubic (fcc) symmetry). In addition, the material is readily deformed and the structure reduced in size.

This means that we have a very important observation which actually applies to both systems. The structures that we are nearly always working with are non-equilibrium forms. At first sight, this would seem to be a major problem when we are characterizing the macroscopic behaviour. It is not the case, however, as we shall see as we can obtain good predictions of the behaviour in the laboratory by just considering the nearest-neighbour interactions to be the dominant term. However, we must be wary. For example, we may have a weakly flocculated concentrated dispersion which appears to be space-filling – in other words, we do not see sedimentation occurring. However, we may see that many hours later, the structure suddenly compacts and we observe a rapid sedimentation. This is simply the structure locally densifying to reduce the local energy. This can only occur at the expense of supports elsewhere, as at some point the 'overburden' becomes critical as the structure becomes progressively weaker in some regions concomitant with densification in other regions. We then observe a sudden collapse. This represents an important problem in designing non-sedimenting dispersions which may be subjected to long-term storage.

There is one other structure which we should consider at this point. This is the type of structure that is formed by particles with a strong interparticle attraction, that is, in coagulated dispersions. If we carry out computer simulations, or careful experiments in the laboratory, these diffusion-limited aggregates have a fractal structure. In a concentrated dispersion, the growth of the fractal aggregates means that they interpenetrate. In practical terms, the complexity involved in attempting to describe such situations is rarely worth tackling as the systems that we use are all heavily shear-processed and this dominates the structure. For example, we may coagulate a monodisperse concentrated dispersion, but when sheared we find that we have a system of spherical aggregates in which the particles are in dense random packing [4]. These systems show a marked change, which is irreversible, when first subjected to shear but subsequently change very little and so the initial state is of minor importance to that produced by shear processing.

Returning to the pair distribution function, we have seen how we have a periodic function which gives the concentration of particles at a distance r from the central particle relative to the global average number density. Hence, we can define the number of particles in a shell dr thick from the volume of the shell, the global average density and $g(r)$, as follows:

$$\text{number} = \rho_0 4\pi r^2 g(r) dr \qquad (9.4)$$

Hence, if we integrate this expression over a distance corresponding to the first nearest-neighbour shell, we have the coordination number for the structure, z, as follows:

$$z = 4\pi\rho_0 \int_{2a}^{r_{min}} r^2 g(r) \mathrm{d}r \qquad (9.5)$$

Of course, the number density may be written in terms of the volume fraction of the dispersion and the particle volume, giving:

$$z = \frac{3\varphi}{a^3} \int_{2a}^{r_{min}} r^2 g(r) \mathrm{d}r \qquad (9.6)$$

The results of such a calculation are shown in Figure 9.3 for the two situations illustrated in Figure 9.2, namely a weakly flocculated system and a strong repulsive system with long-range interactions. The coordination numbers obtained for the weakly attractive system [5] correspond to a latex in 'high salt' but sterically stabilized with non-ionic surfactant. The pair potential had an attractive minimum of $7k_B T$ at $\sim 10\,\mathrm{nm}$ from the surface. The shape of the curve is sigmoidal and as the volume fraction is increased we see that the coordination number increases most rapidly at $\varphi \sim 0.3$. What is happening in this structure is that the coordination number is increasing while the particles are at close separation, i.e. in the potential minimum. On a practical application note, it is common to encourage the separation of particles by causing some aggregation and then centrifuging or using a filter press. When the coordination number increases rapidly, the structure becomes very strong and 'dewatering' becomes difficult. We noted that as $Q \rightarrow 0$, $S(Q)$ gives the osmotic

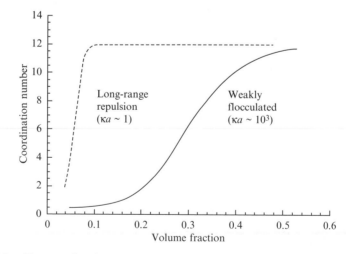

Figure 9.3. The coordination numbers calculated for the structures formed from both weakly flocculated and those with long-range repulsion between the particles.

osmotic compressibility. This is the excess osmotic pressure, i.e. that due to the particle–particle interactions. Thus, from Equation (9.2a) we can see immediately that we expect the compressibilty to be reduced as $g(r)$ increases. Figure 9.3 also shows the equivalent result for a system with long-range repulsion between the particles. The $\kappa a \sim 1$ condition means that the particles become highly ordered at a low volume fraction. When such a system is concentrated, the structure (fcc) is the same short-range structure throughout most of the concentration range but the separation is changing, and qualitatively quite different from the weakly attractive system.

A consequence of the structure that we see developing as we concentrate dispersions is that *the compressibility of the structure* is reduced. This means that the structure resists changes in volume as we might expect to see in a centrifugation or filtration experiment. We are talking about the bulk elastic modulus of the structure so that we are observing solid properties with the concentrated dispersion. This resistance to a change in the shape of the structure also shows in shear or extension experiments and the resistance is a function of the $g(r)$ and the rate of change of the interparticle force with changes in separation. However, we also find that under high stress, or sometimes under low stress applied for long times, the systems flow. This area of study is known as *rheology* and is a key feature of most of the concentrated dispersions that we use everyday.

3 RHEOLOGY

3.1 Definitions

When we study the deformation and flow of concentrated systems, we must define the forces and deformations carefully. The stress is defined as the force per unit area over which it is applied and so has the units of pascals (SI unit of pressure and stress). The strain is the deformation relative to the original dimension, and so is dimensionless. An arbitrary applied stress which results in a deformation has to be described in all three dimensions. There are shear stresses as well as stresses normal to the reference planes in the material. To describe this, we need to resort to tensor algebra and this has operational rules for the manipulation of the equations which may not be too familiar to many of those interested in colloids. The complexity is avoided in the laboratory by carefully controlling the way in which the materials deform or the way the stress is applied. Thus, we can limit our mathematical manipulations to simple linear algebra and a little complex number algebra in the contents of this present chapter. The symbol commonly used for stress in much of the rheological texts is σ, and that used for the strain is γ, and we will use these here, despite the risk of confusion due to their use to also denote charge and

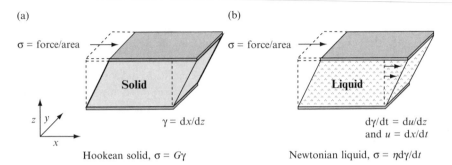

(a)　　　　　　　　　　　　　　　(b)

σ = force/area　　　　　　　　　　σ = force/area

Solid　　　　　　　　　　　　　　Liquid

z y 　　　$\gamma = dx/dz$　　　　　$d\gamma/dt = du/dz$
x　　　　　　　　　　　　　　　and $u = dx/dt$

Hookean solid, $\sigma = G\gamma$　　　Newtonian liquid, $\sigma = \eta \, d\gamma/dt$

Figure 9.4. Definitions of the shear stress, σ, the shear strain, γ, and the rate of shear, $\dot{\gamma}$, for (a) a Hookean solid, and (b) a Newtonian liquid.

surface tension. Moreover, it will implied that they are for shear stresses and strains, as shown in Figure 9.4.

In addition to stress and strain, the other parameter that we need to define is the timescale of the experiment. It is important to relate this to a characteristic timescale of the material behaviour. This is the type of thing that we do automatically when, for example, we carry out an experiment where a chemical change is taking place in a wide range of situations such as working with radioisotopes or when simply observing an expiration date on a commercial product, such as a foodstuff, in our everyday lives. In a rheological experiment, we are considering movement – the timescale that is relevant here is that of the diffusional process. It is the slowest process which is rate-determining and with a condensed phase we need to define a long-time self-diffusion time. The Einstein–Smoluchowski equation gives the average time for a particle to diffuse through one particle radius and by using the Stokes–Einstein equation which defines the diffusion coefficient, we have the characteristic time, τ, as follows:

$$\tau = \frac{a^2}{D} = \frac{6\pi\eta(0)a^3}{k_B T} \tag{9.7}$$

The important point to note in Equation (9.7) is that the limiting viscosity is that of the colloidal dispersion under low-shear-rate conditions and not that of the continuous phase, η_0. This is the *effective medium* concept which gives a first-order estimate of all of the multi-body hydrodynamic interactions. If the experimental time is t_{ex}, we can define the dimensionless group, known as the *Deborah number*, as follows:

$$D_e = \frac{\tau}{t_{ex}} \tag{9.8}$$

If a material is only deformed by an applied stress for such a short time that the particles do not have time to diffuse very much, when the stress is removed the original shape is regained and the deformation is termed *elastic*. The reason is that as the structure formed by the particles is deformed, work is done and this is stored in that structure. Removal of the stress allows the structure to move back to a lower-energy configuration. The situation where the experiment continues for a time which is much longer than the time for particle diffusion results in a permanent or viscous deformation of the material as the structure moves towards a low-energy configuration in the new shape. We say that the stress has relaxed and the characteristic time, τ, is thus the *stress relaxation time*. There are three ranges of the Deborah number that can be identified, as follows:

$$D_e \ll 1 \qquad D_e \sim O(1) \qquad D_e \gg 1$$
liquid-like   viscoelastic   solid-like

and it is immediately clear that there is progression from the behaviour of liquids through to that of solids. This appears to contradict our normal experience where the liquid/solid transition is very sharp but this only means that the relaxation time changes dramatically with a very small change in temperature, concentration or pressure (stress). With concentrated colloids, surfactants and polymer solutions, the change in relaxation time is not so sudden as the structural components can move over larger distances and we can often make use of the broader viscoelastic range to produce the correct handling properties for colloidal products.

It is important to note that experimental timescales are not arbitrarily chosen by laboratory instrumentation. Such instruments are built with timescales that either simulate a process that we require numerical data for, such as the brushing of a paint film or the sedimentation of a colloidal dispersion on storage, or to quantify an observation that we have made. This latter aspect takes the problem to our in-built 'bio-timescale'. We are aware of changes taking place over times ranging approximately from 10^{-3} to 10^3 s. Hence, if the relaxation time lies within the range 10^{-4} s $< \tau < 10^4$ s, the viscoelastic responses will be observed as we handle the material.

Many materials are produced as colloidal dispersions because of their liquid-like behaviour under at least certain conditions. Often, this is under high stresses and/or high strains and the property that is required is the 'correct' viscosity. In practice, the last stage in a formulation is often the adjustment of the concentration by the addition of solvent to a particular viscosity value. The characteristic timescale for a continuous shear process is just the reciprocal of the shear rate, $\dot{\gamma}^{-1} = dt/d\gamma$, which is the time for unit strain to occur. As this is the characteristic experimental timescale, this immediately suggests the dimensionless group which is known as the *Péclet number*, P_e, which for a concentrated colloid is given by the following:

$$P_e = \tau\dot{\gamma} = \frac{6\pi\eta(0)a^3\dot{\gamma}}{k_B T} \tag{9.9}$$

Note that again we are using the suspension viscosity limit at low shear rate as the viscosity of the effective medium. We should explore this in a little detail. At low Péclet numbers, the relaxation time is short when compared to the characteristic shear time and the structure is relatively unperturbed by the shear action. However, at high Péclet numbers the shear rate is dominant and the structure is controlled by the hydrodynamic forces as the particles have insufficient time to diffuse to allow for the shape change. This means that at $P_e \sim 1$ the structure should be intermediate. We will explore how this should effect the viscosity in the next section while we note at this stage that the viscosity must be a function of the structure.

3.2   The Viscosities of Concentrated Dispersions

In Chapter 8, the viscosity of a dilute dispersion was presented and the equation describing the variation with volume fraction was based on rigorous hydrodynamic analysis, and in the dilute limit the result due to Einstein was as follows:

$$\eta = \eta_0(1 + [\eta]\varphi + \dots) \tag{9.10}$$

where the intrinsic viscosity, $[\eta] = 2.5$ for hard spheres at a volume fraction of φ. As rigorous hydrodynamics cannot take the result much above $\varphi \sim 0.5$, the effective medium approach due to Krieger and Dougherty [6] will be used. The most straightforward analysis was presented by Ball and Richmond [7] and is repeated here as it emphasizes some important points. The following equation gives us the rate of increase of viscosity with volume fraction as:

$$\frac{d\eta}{d\varphi} = [\eta]\eta_0 \tag{9.11}$$

Now, when we consider the small replacement of a volume of the continuous phase by some further particles we may expect a similar rate of change in viscosity, and so:

$$d\eta = [\eta]\eta(\varphi)d\varphi \tag{9.12}$$

Here, we are using $\eta(\varphi)$ as the viscosity of the system that the newly added particles will 'experience'. As the new particles have to have been added to a constant-volume system, the change in concentration must be corrected to the available volume:

$$\frac{d\eta}{\eta(\varphi)} = [\eta]\frac{d\varphi}{\left(1 - \dfrac{\varphi}{\varphi_m}\right)} \tag{9.13}$$

Here, the maximum concentration at which flow can occur is φ_m and so φ/φ_m is the excluded volume of the dispersion, i.e. $(1 - \varphi/\varphi_m)$ is the liquid volume that could be replaced by more particles. Integration through the volume fraction range, with the boundary condition that as $\varphi \to 0$, $\eta(\varphi) \to \eta_0$, gives the *Krieger–Dougherty equation* for the volume-fraction dependence of the viscosity as:

$$\frac{\eta(\varphi)}{\eta_0} = \left(1 - \frac{\varphi}{\varphi_m}\right)^{-[\eta]\varphi_m} \tag{9.14}$$

For a monodisperse system of hard spheres, $[\eta] = 2.5$. The value of φ_m varies [5] from the value of the liquid–solid transition for hard spheres under quiescent conditions (i.e. 0.495 for freezing and 0.54 for melting) to 0.605, corresponding to the flow of hexagonally packed layers at high shear rates. So, we can expect the viscosity to be a function of the shear rate. These values give, for uniform hard spheres, the limiting viscosities at high and low shear rates as follows:

$$\frac{\eta(0)}{\eta_0} = \left(1 - \frac{\varphi}{0.5}\right)^{-1.25}; \quad \frac{\eta(\infty)}{\eta_0}\left(1 - \frac{\varphi}{0.605}\right)^{-1.51} \tag{9.15}$$

We have used $\varphi_m = 0.5$ as the maximum volume fraction in the lower shear limit but we should recognize that the liquid and solid phases can co-exist between volume fractions of 0.495 and 0.54 and this will allow some flow to occur between these limits. The shear-thinning behaviour is illustrated in Figure 9.5 in which the high and low shear limits of the viscosity are plotted as a function of the volume fraction of the dispersion.

There are four regions shown in Figure 9.5. In Region A, the dispersions behave as simple Newtonian fluids with no discernible shear dependence, while in Region B some shear-thinning may be observed. Throughout Region C, viscoelastic liquid-like behaviour can be found with both the high-shear-limiting viscosity, $\eta(\infty)$, and low-shear-Newtonian limit, $\eta(0)$, being accessible. The concentrated dispersion in this region can be characterized as a weak gel as significant elastic behaviour can be observed. We can also call a colloid in this region a complex fluid. In Region D, the behaviour is that of a viscoelastic solid so that $\eta(0)$ is no longer accessible but there is a yield stress

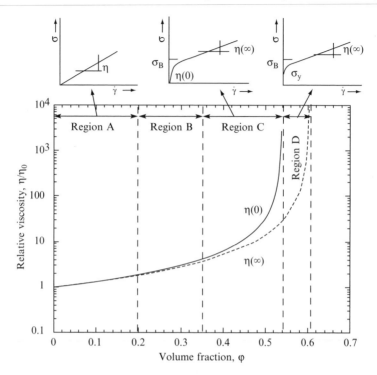

Figure 9.5. The relative viscosity as a function of volume fraction in the high- and low-shear limits. The dispersion is Newtonian in Region A, 'pseudoplastic' in Region C and viscoelastic ('plastic') in Region D (some shear-thinning can be observed in Region B).

which when exceeded, the structure will melt and flow occurs. This is the *soft solid* stage. At high shear stresses, a value of $\eta(\infty)$ can be measured. At the upper limit of this region, the material will fracture under a sufficiently high stress and we will not observe flow. Table 9.1 summarizes the measurable parameters which are characteristic of the material. Figure 9.5 also shows the type of shear stress–shear rate curves that are measured, and so in Region A the linear response of a Newtonian fluid is seen, Region C displays typical 'pseudoplastic' behaviour and Region D is characteristic of a plastic material. The static and dynamic yield stresses are shown where appropriate.

The shear-thinning response of the pseudoplastic dispersion is a smooth change from the low-shear Newtonian behaviour to the high-shear plateau. The curve is well described by the *Krieger equation* [8] which is written in terms of the reduced stress, σ_r:

$$\eta(\sigma_r) = \eta(\infty) + \frac{\eta(0) - \eta(\infty)}{1 + (b\sigma_r)^n} \tag{9.16}$$

Table 9.1. The characteristic parameters for a viscoelastic colloid, where $G(\infty)$ is the high-frequency limiting shear modulus, σ_y the static yield value, σ_B the dynamic or Bingham yield value, ω the frequency of an applied small-amplitude shear oscillation, $\dot{\gamma}$ the shear rate and τ the characteristic time

Region	Linear responses – small stresses and/or strains[a]		Non-linear responses – large stresses or strains[b]
Liquid-like	$G(\infty)$ (Pa)	$\eta(0)$ (Pa s)	$\eta(0)$; $\dot{\gamma}^{-1} \gg \tau$ (s)
Region C	$\omega^{-1} \gg \tau$ (s)		$\eta(\infty)$; $\dot{\gamma}^{-1} \ll \tau$ (s)
Solid-like Region D	$G(\infty)$ (Pa)	σ_y (Pa)	$\eta(\infty)$; $\dot{\gamma}^{-1} \ll \tau$ (s)
	$\omega^{-1} \ll \tau$ (s)		σ_B

[a] For example, oscillations.
[b] For example, continuous shear.

The reduced stress, which was derived from dimensional analysis, is directly related to the *Péclet number*:

$$\sigma_r = \frac{a^3 \sigma}{k_B T} = \frac{P_e}{6\pi} \qquad (9.17)$$

For a monodisperse system, $n = 1$, and using a value of the exponent $n > 1$, the viscosity shear-thins over a wider range of stresses than the monodisperse system. The value of b is obtained from the mid-point of the curve. Recalling that P_e is the ratio of the convective to thermal timescales for the particle motion, means that midway between the Brownian-motion-dominated structure (the zero-shear plateau) and the shear-dominated structure (the high-shear plateau) $b\sigma_r = 1$, so that $b = 1/\sigma_{rc}$. The latter parameter, the critical reduced stress, is the value of the reduced stress at the mid-point of the viscosity curve. It is instructive to think of this in terms of the characteristic time τ for the relaxation of the structure. This concept suggests that a spectral range of relaxation times should be included. For example, as the size distribution broadens, the range of relaxation times also broadens because the diffusive motion is a function of particle size and local concentration (as well as pair-potential for particles other than hard spheres). As a starting point, we can sum the contributions of each component to the stress weighted for its probability, p_i, and then we may rewrite Equation (9.16) as:

$$\frac{\eta(\sigma_r) - \eta(\infty)}{\eta(0) - \eta(\infty)} = \sum_i \left(p_i \frac{1}{1 + \sigma_{rci}} \right) \qquad (9.18)$$

where σ_{rci} is the critical stress for the ith component. Figure 9.6 illustrates the response for a system with a single relaxation time and how an example

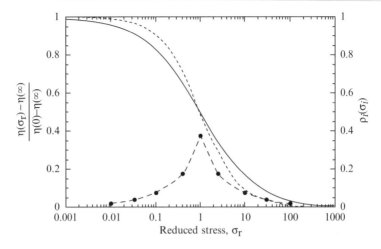

Figure 9.6.   The effect of a range of relaxation times on the shear-thinning response of a dispersion: (- - - -) single relaxation time; (——) spectral range; (- ● -) spectrum.

spectrum over a range of different times broadens the response from ca. three or four orders of magnitude to at least six. We should keep in mind that for a monodisperse system, we expect a spectrum of relaxation times from the fluctuations of the local concentration due to the collective diffusion of the particles [5]. We will see later how this concept of using a spectrum of relaxation times fits naturally with the practice adopted in analysing linear viscoelastic data. This is most appropriate when used with polymer solutions where we can usually identify the low-shear viscosity with the low-frequency dynamic viscosity (the *Cox–Mertz rule*). With particulate systems, this correspondence is not always good but nevertheless it is an important characterization.

As the colloidal forces between the particles become significant, the stress required to move them relative to each other increases. This means that the total stress for the flowing system is the sum of the hydrodynamic and the colloidal terms. Note that for a simple hard-sphere system, the colloidal term is the Brownian term. At very low shear rates, the hydrodynamic term becomes much smaller than the colloidal contribution. The result is that the zero-shear viscosity is controlled by the colloidal forces. There are statistical mechanical models available [9] to calculate the viscosity which integrate the contributions from the colloidal forces over the structure. The latter is given by the pair distribution function, which is distorted by the shear field. For the case of charge-stabilized dispersions, we may treat the particles as 'effective' hard spheres where the 'effective' radius for such spheres is given in the low-shear limit as the collision radius of the particles during a Brownian motion encounter. This was carried out by equating the electrostatic repulsive force with the thermal force. Russel *et al.* [10] derived the closest distance between particle centres, which is, of course, the value of the 'effective' hard sphere diameter, $r_0(0)$:

$$r_0(0) \approx \kappa^{-1} \ln \{\alpha / \ln [\alpha / \ln (\alpha / \ln \ldots)]\}$$

$$\text{where } \alpha = \left[\frac{4\pi\epsilon\kappa(a\zeta)^2 \exp(2\kappa a)}{k_B T} \right] \tag{9.19}$$

This is just an increase in the excluded volume of the particle and so modifies the value of the maximum volume fraction at which the viscosity diverges. Hence, the low-shear limiting viscosity for charge-stabilized dispersions now becomes:

$$\eta(0) = \eta_0 \left(1 - \frac{\varphi}{\varphi'_m} \right)^{\frac{5\varphi'_m}{2}}$$

$$\text{where } \varphi'_m = 0.495 \left(\frac{2a}{r_0(0)} \right)^3 \tag{9.20}$$

Figure 9.7 shows how sensitive the low-shear viscosity is to particle size for small charged particles at moderate electrolyte concentrations. Here, the calculation used values of $\zeta = 50\,\text{mV}$ with a charge density of the Stern layer of

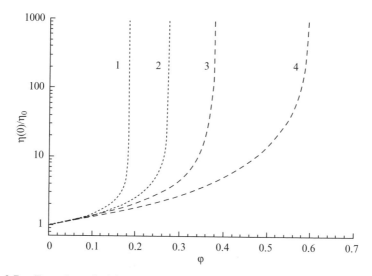

Figure 9.7. Zero-shear limiting viscosity as a function of particle radius: $\zeta = 50\,\text{mV}$; $\sigma_\delta = 1\,\text{mC cm}^{-2}$: (1) $a = 50\,\text{nm}$; (2) $a = 100\,\text{nm}$; (3) $a = 250\,\text{nm}$; (4) high-shear limit.

$\sigma_\delta = 1\,\mu\mathrm{C\,cm}^{-2}$, with the latter quantity being used to calculate the value for $\kappa(\varphi)$ as this gives the number of counter-ions in the diffuse layer. Equation (9.19) gave a value of $r_0(0) \sim (2a + 5/\kappa)$, which serves to illustrate the importance of the slow decay with distance of the pair-potential and indicates the origin of the large excluded volume of small charged particles. We should also note that the very steep rise in the viscosity as the liquid/solid transition is approached will be less sharp in practice due to the co-existence of the two phases in the volume fraction range 0.50–0.54.

Similar results are found for particles with attractive forces dominating the long-range part of the pair-potential, resulting in a weakly flocculated system. The modelling requires a different approach and the particle size dependence is different. In most practical cases, colloidal formulations have many components and so it is not always clear how the behaviour relates to the theoretical calculations which are frequently carried out for model dispersions of monodisperse spheres. However, there are some common origins so that we may formulate some general observations.

3.3 Viscosity Summary

(1) There are three forces which govern the stress that we observe when we measure the viscosity as a function of shear rate, namely hydrodynamic forces, Brownian (thermal) forces and colloidal forces arising from the form of the pair-potential.

(2) In the high-shear limit, the dispersion structure and measured stress is dominated by the hydrodynamic forces.

(3) The high-shear limit of φ_m is a function of the nature of the disperse phase. Polydispersity increases the value [5], as do the fluid particles due to particle deformation, although this does not occur until high volume fractions are reached as the surface forces are strong for particles with colloidal dimensions. Particle anisometry increases this value.

(4) We should note that in many applications much higher shear rates occur than can be accessed by using most laboratory rheometers and also there is frequently an extensional component present. With particulate systems, this does not present a problem as the shear-thinning occurs over a relatively narrow range of shear rates or applied stresses and so the high-shear limit can be reached, or at least approached. However, this is not the case with solution polymers which have a much broader response due to first, the broad molecular weight distributions, which are much broader than the particle size distribution produced with most particulates. Secondly, the relaxation modes of each chain varies from the timescales ranging from the centre-of-mass diffusional time of the whole molecule down to that of a single segment.

(5) In the low-shear limit, the dispersion structure and excluded volume of the particles are governed by the balance of the Brownian and colloidal forces. Measurements under low-shear conditions are therefore particularly useful in identifying changes in colloidal interactions as formulation conditions are varied.

(6) The prediction of the shear-thinning behaviour is usually not possible as the complexity of most practical systems rules this out and we have to rely on experimental measurements.

(7) Many formulations are produced in the concentration regime between where the low-shear viscosity increases rapidly and the high-shear limit. In this region, it is usually the viscoelasticity that is most important as this dominates our perception of the 'handling properties' of the dispersion, whereas the extremes of behaviour are the most important parameters for the applications. For example, brushing, spreading or spraying are all high-shear processes with shear rates $\dot{\gamma} \geq 10^3$ s^{-1}, while sedimentation is a 'long-time' low-stress process where $\dot{\gamma} \to 0$.

4 LINEAR VISCOELASTICITY OF COLLOIDAL DISPERSIONS

The discussion of viscoelasticity in this text is restricted to linear responses, where the viscosity is Newtonian and the elasticity obeys *Hooke's law*, and so the limiting constitutive equations for shear are:

$$\sigma = \eta\dot{\gamma}; \quad \sigma = G\gamma \tag{9.21}$$

This means that in the laboratory we have to restrict our measurements to low stresses and strains, but moreover, we must check that this is the case if we are to use the analysis of the data to give characteristic parameters of the dispersion rheology. Hence, we should determine that if the stress or strain is changed then the corresponding strain or stress changes in proportion.

4.1 Constitutive Equations

When we make measurements in the laboratory, it is convenient to fit the data to a curve so that we may summarize the data in terms of as few characteristic parameters as will give an adequate description of our observations. This serves two purposes. It enables us to make a quantitative comparison of different materials, which could be new batches of the same formulation or different formulations. Thus, the subjectivity is minimized. In addition, we may be able to interpret the characteristic parameters in terms of the interactions between the microstructural components. Even if the latter becomes difficult, the former is an imperative.

When we devise experiments to determine the viscoelastic response of a material, we need to vary the stresses or strains over a range of timescales to explore the response over a range of Deborah numbers. For example, we can apply a step stress/strain and follow the response over time or we may oscillate the material over a range of frequency. The measured data are in terms of the elastic modulus if we measure the stress as a function of applied strain, or the compliance if the strain is recorded as a function of the applied stress. The problem that we then have is to fit the experimental curves. As an aid to this end, we can invoke the responses of mechanical analogues to help us to derive suitable equations. Note that it is also possible to use the responses of electrical circuits for the same purpose. The algebra is similar in both cases. The utility of the constitutive equations is primarily to give the correct response of the material over the range studied and, if possible, to yield the limiting behaviour.

The mechanical analogues use Hookean springs for the elastic behaviour and Newtonian dashpots or dampers to give the viscous response. These can be coupled in many combinations to give mechanical responses similar to the experimental curves that we are attempting to simulate. A few of these are shown in Figure 9.8, along with the corresponding equations relating the

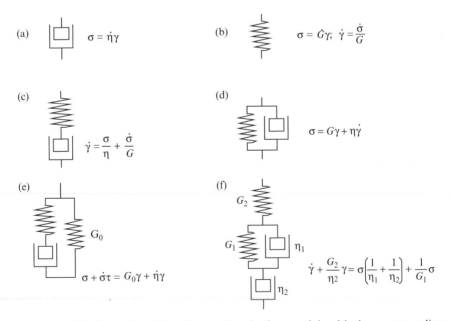

(a) $\sigma = \dot{\eta}\gamma$

(b) $\sigma = G\gamma; \quad \dot{\gamma} = \dfrac{\dot{\sigma}}{G}$

(c) $\dot{\gamma} = \dfrac{\sigma}{\eta} + \dfrac{\dot{\sigma}}{G}$

(d) $\sigma = G\gamma + \eta\dot{\gamma}$

(e) $\sigma + \dot{\sigma}\tau = G_0\gamma + \dot{\eta}\gamma$

(f) G_2, G_1, η_1, η_2 $\quad \dot{\gamma} + \dfrac{G_2}{\eta_2}\gamma = \sigma\left(\dfrac{1}{\eta_1} + \dfrac{1}{\eta_2}\right) + \dfrac{1}{G_1}\sigma$

Figure 9.8. Mechanical models of linear viscoelastic materials with the corresponding equations. The characteristic time for all models is $\tau = \eta/G$.

stress and strain. Which ones we make use of depends on the form of the experiment that we carry out. The appropriate equation is then used to calculate the form of the modulus or compliance of *the model* for the appropriate form of the temporal dependence of the applied strain or stress. Although it would be possible to construct very complicated analogues, these would be of little utility as we loose uniqueness of the fit if we have many 'characteristic parameters'. We will therefore restrict our use to just a few of the possible models, keeping in mind that that the object is to compare and record the responses of materials with as few characteristic parameters as possible. A quantitative microstructural interpretation of these may occasionally be possible for the simplest systems.

5 PHENOMENOLOGY

This section will describe the various experiments that are carried out to characterize linear viscoelasticity. The first method described – the application of an oscillating deformation – is one frequently used and the derivation of the responses from a mechanical analogue is illustrated in some of these as examples. The results of the algebraic analyses for the other types of experiments are only simply stated.

5.1 Oscillating Strain

In this experiment, the concentrated colloid is usually subjected to a sinusoidally oscillating strain and the resulting stress is measured at a variety of frequencies. During the experiment the rheometer records three parameters:

- the peak strain that is applied
- the resulting peak stress
- the difference in phase between the strain and stress wave forms at each frequency

Figure 9.9 illustrates the form of the stress that we would measure for a material subjected to an oscillating strain, which we will assume to have a frequency of 1 rad s^{-1}. In this example, the maximum strain, γ_0, is 0.05 and the maximum stress σ_0 is 25 Pa. Now, the stress per unit strain is the modulus at an oscillation frequency of ω, as follows:

$$G^*(\omega) = \frac{\sigma_0}{\gamma_0} \qquad (9.22)$$

which for the example data shown in Figure 9.9 gives a value of 0.5 kPa for the *complex modulus*, G^*. This modulus has contributions from both a storage

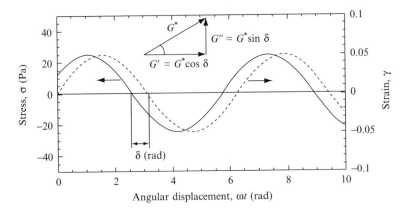

Figure 9.9. Forced oscillation experiment for a material with $G^* = 0.5\,\text{kPa}$ at a strain of 5 %: $\delta = 30°$; $G' = 0.433\,\text{kPa}$; $G'' = 0.25\,\text{kPa}$.

(elastic) and a loss (viscous dissipation) term. If the material was a simple Hookean solid, the position of the maximum stress would coincide with that of the maximum strain. On the other hand, if the experiment was carried out on a Newtonian liquid, the maximum stress would be coincident with the maximum rate of strain. The phase shift, δ, enables us to assign values to the storage and loss contributions to the complex modulus, as follows:

$$G^*(\omega) = G'(\omega) + iG''(\omega)$$
$$G'(\omega) = G^*(\omega)\cos\,\delta$$
$$G''(\omega) = G^*(\omega)\sin\,\delta \tag{9.23}$$

Here, we are using the normal complex number algebraic notation where $i^2 = -1$ and :

$$G'(\omega) + iG''(\omega) = G^*(\omega)(\cos\,\omega t + i\sin\,\omega t) = G^*(\omega)e^{i\omega t} \tag{9.24}$$

Alternatively, the complex viscosity can be defined as follows:

$$\eta^*(\omega) = \frac{G^*(\omega)}{\omega} = \eta'(\omega) + i\eta(\omega) \tag{9.25}$$

and the dynamic viscosity as the real, or in-phase, part of the complex viscosity:

$$\eta'(\omega) = \frac{G''(\omega)}{\omega} \tag{9.26}$$

Returning to the curves in Figure 9.9, the values for the storage and loss moduli, respectively, are $G'(\omega) = 0.433\,\text{kPa}$ and $G''(\omega) = 0.25\,\text{kPa}$ at 1 rad s^{-1}. Experimentally, we determine the moduli over as wide a range of frequency as is practicable, but the problem is then to condense these data to as few characteristic material constants as will provide an accurate fit to such data.

To illustrate the process further, we can take the constitutive equation for a *Maxwell model*, which has a single spring in series with a dashpot, as shown in Figure 9.8(c). So, for an oscillating strain we have:

$$\dot{\gamma}^* = \frac{\dot{\sigma}^*}{G} + \frac{\sigma^*}{\eta} \tag{9.27}$$

and the complex stresses and strains are written in terms of the peak stresses and strains that are measured:

$$\gamma^* = \gamma_0 \exp\ (i\omega t); \ \dot{\gamma}^* = i\omega \exp\ (i\omega t) = i\omega\gamma^*$$
$$\sigma^* = \sigma_0 \exp\ [i(\omega t + \delta)]; \ \dot{\sigma}^* = i\omega\sigma_0 \exp\ [i(\omega t + \delta)] = i\omega\sigma^* \tag{9.28}$$

Substitution of the above into Equation (9.27) and rearranging gives:

$$\frac{i\omega\gamma^*}{i\omega\sigma^*} = \frac{1}{G^*(\omega)} = \frac{1}{G} + \frac{1}{i\omega\eta} \tag{9.29}$$

Noting that the characteristic or time of the material is $\tau_r = \eta$ (Pa s)/G (Pa), we can then rearrange Equation (9.29) to give:

$$G^*(\omega) = G\left(\frac{i\omega\tau_r}{1 + i\omega\tau_r}\right) \tag{9.30}$$

We can separate the complex modulus into the strorage and loss moduli and multiplying Equation (9.30) throughout by $(1 - i\omega\tau_r)$ gives:

$$G'(\omega) + iG''(\omega) = G\left[\frac{(\omega\tau_r)^2}{1 + (\omega\tau_r)^2}\right] + iG\left[\frac{(\omega\tau_r)}{1 + (\omega\tau_r)^2}\right] \tag{9.31}$$

If for our example we assume the relaxation time to be 0.5 s, the value of G from the data given can be readily calculated, as follows:

$$G''(1) = G\left(\frac{0.5}{1 + 0.5^2}\right), \text{ which gives } G = 0.625\,\text{kPa}$$

$$\text{and } \eta = 0.5 \times 0.626 = 0.318\,\text{kPa s}$$

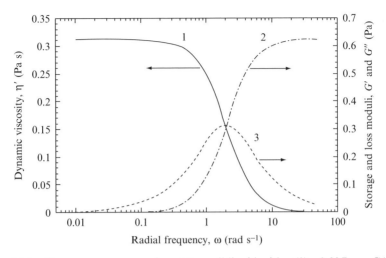

Figure 9.10. Frequency responses for a Maxwell liquid with $\eta(0)=0.32$ Pa s, $G(\infty)=$ 0.63 Pa and $\tau_M = 0.5$ s: (1) dynamic viscosity; (2) storage modulus; (3) loss modulus.

The frequency dependences of the storage and loss moduli are as illustrated in Figure 9.10. Features of the curves that are typical of a Maxwell model are that:

- the storage modulus reaches a plateau at high frequency, $G'(\omega) \rightarrow G(\infty)_{\omega \rightarrow \infty}$, noting that $G(\infty) = G$ in the model;
- $G''(\omega) > G'(\omega)$ when $\omega < 1/\tau_r$, but $G''(\omega) < G'(\omega)$ when $\omega > 1/\tau_r$;
- $G''(\omega)_{\omega = 1/\tau_r} = 0.5G(\infty)$;
- at low frequencies, the dynamic viscosity reaches a plateau, and so $\eta'(\omega)_{\omega \rightarrow 0} \rightarrow \eta(0)$, noting that $\eta(0) = \eta$ in the model.

At this point, it is useful to consider some experimental data obtained using a colloidal product. The material in this case was a shower gel. The formulation consisted of a concentrated surfactant system at a high enough concentration so that a condensed phase of 'worm-like' micelles can form. The rheology of the formulation had been 'fine-tuned' with a little polymer. The experimental data obtained from this system when using an applied strain of 10% are plotted in Figure 9.11. The data are shown by symbols, with the corresponding curves being calculated for a Maxwell model. The value for the relaxation time was 0.1 s and for $\eta(0)$ 30 Pa s. Both were easily determined from the curves and these values gave $G(\infty) = \eta(0)/\tau_r = 300$ Pa. Note that the fits become poor at frequencies above > 10 Hz. This is caused by the applied frequency approaching a resonance frequency of the measuring unit. However,

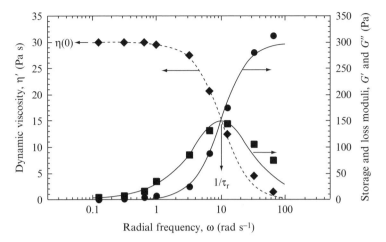

Figure 9.11. Oscillation responses for a sample of shower gel. The symbols represent experimental data, with the curves calculated for a Maxwell fluid with $\eta' = 30\,\text{Pa s}$ and $\tau = 0.1\,\text{s}$.

it is evident from the forms of the rest of the plots that the shower gel is a viscoelastic fluid and a Maxwell model provides an adequate description of the material with a well-defined characteristic time and a clear limiting viscosity at low frequencies. Hence, these are the only parameters which we need to record to characterize the material and enable a batch-to-batch comparison.

The next example that we are going to consider is a household cleaner used for kitchens. Like the shower gel, this contained a large amount of surfactant in order to remove greasy deposits. However, it also contained an abrasive powder to aid the cleaning action. This abrasive was calcium carbonate and the particle size was in the upper part of the colloidal range, that is, $1\,\mu\text{m} <$ mean diameter $< 10\,\mu\text{m}$. If the particle size is too large, the material feels 'gritty' and not creamy; if too small, the abrasive action is sacrificed. Now, the density difference between the calcium carbonate and the surfactant concentrate is $\sim 2000\,\text{kg m}^{-3}$. If this was the final formulation, the particles would sediment slowly as the system would be a viscoelastic fluid like the shower gel. However, poly(acrylic acid) was also included as a 'rheological modifier'. Divalent Ca^{2+} ions bind strongly to $-COO^-$ groups and cross-link the polymer. This cross-linking makes such a system a viscoelastic solid and sedimentation is prevented as the limiting viscosity at low stresses, $\eta(0) \to \infty$. Careful choice of the polymer concentration limits the cross-link density and the structure is readily strain 'melted' (i.e. broken). The experimental data are presented in Figure 9.12, with the curves representing the values calculated for the mechanical analogue shown in Figure 9.8(e), that is, a Maxwell model in parallel with a single spring providing the solid-like response at long

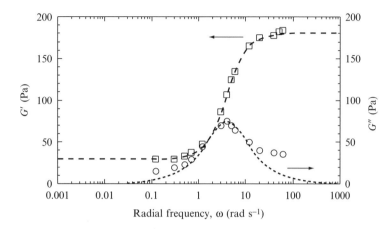

Figure 9.12. Oscillation responses for a sample of household cleaner: □, storage moduli; ○, loss moduli. The curves are plotted employing the mathematical model shown in Figure 9.8(e), using the following parameters: $G(0) = 30$ Pa; $\tau_1 = 0.25$ s; $G_1 = 150$ Pa (giving $G(\infty) = 180$ Pa).

periods of time. It is important to recognize that the presence of a $G(0)$ value with the curves of the storage and loss moduli diverging is indicative of solid-like behaviour and clear control of sedimentation. Such curves need three parameters, and the values used were: $G(\infty) = 180$ Pa, $G(0) = 30$ Pa and $\tau_1 = 0.25$ s ($\eta(0)$ is, of course, ∞). In terms of the model shown in Figure 9.8(e), this gives $G_0 = 30$ Pa, $G_1 = (180 - 30) = 150$ Pa and $\eta_1 = (150 \times 0.25) = 37.5$ Pa s. The fit is moderately reasonable in terms of the storage modulus. However, the loss modulus is poorly described at frequencies both higher and lower than the characteristic frequency. This broadening of the response is indicative of there being more than one relaxation process present – not a surprising result for a system consisting of 'worm-like' micelles, cross-linked polymer chains and large particles of colloidal size. We get a much better fit to the experimental data when we use a more complicated model. A good fit was obtained by using a four-element model consisting of three Maxwell models and a spring in parallel. The values for each of the components are given on Figure 9.13. Note that the main peak is barely changed and that a smaller $G(0)$ is predicted but it is still of the same order of magnitude as the simpler model. For routine quality control purposes, the data obtained from the simpler approach illustrated in Figure 9.12 would often be adequate. This may not always be the case though and in terms of product development it would be important to determine, numerically, how the values shown in Figure 9.13 vary with component changes. Of course, the data range shown in this figure should be broadened as much as possible but the general concept introduced here is an important one.

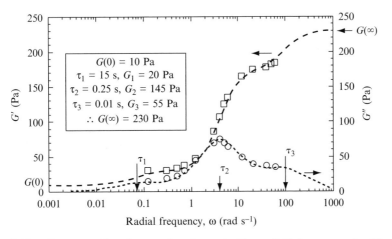

Figure 9.13. Corresponding data to that shown in Figure 9.12 for a sample of household cleaner: □, storage moduli; O, loss moduli. The curves are plotted in this case by using a four-element mathematical model consisting of three Maxwell models and a spring in parallel, giving a much improved fit to the experimental data.

We are seeing here a simple illustration of a spectral response, i.e. there is a series of processes occurring which all contribute to the overall response. Mathematically such a series is described as the *generalized Maxwell model*, (see Figure 9.14) which is represented as follows:

$$G'(\omega) = G(0) + \sum_1^n G(\infty)_n \frac{(\omega\tau_n)^2}{1 + (\omega\tau_n)^2}$$

$$G''(\omega) = \sum_1^n G(\infty)_n \frac{(\omega\tau_n)}{1 + (\omega\tau_n)^2} \qquad (9.32)$$

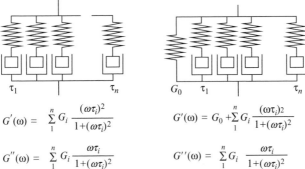

Figure 9.14. Illustrations of generalized Maxwell models for (a) a viscoelastic fluid, and (b) a viscoelastic solid.

It is very important to recognize that this is qualitatively different from the viscous response that we discussed earlier. In that case, we were looking at the shear melting of the structure as the rates of deformation became too rapid for parts of the structure to accommodate the change. This was a non-linear measurement. In the case we are now discussing, the measurement is linear and the structure is not being significantly modified by the experiment, that is, this is a non-destructive experiment, while the former one was a destructive experiment. We may be able to fit our data by using Equation (9.32) but we should also recognize that the way in which Equation (9.32) is written implies that each relaxation process has equal weighting. An alternative way of expressing the spectral response is through an integral equation instead of a summation, as follows:

$$G'(\omega) = G(0) + \int_{-\infty}^{+\infty} H \frac{(\omega\tau)^2}{1 + (\omega\tau)^2} d\ln \tau$$

$$G''(\omega) = \int_{-\infty}^{+\infty} H \frac{(\omega\tau)}{1 + (\omega\tau)^2} d\ln \tau \qquad (9.33)$$

where H is known as the *relaxation spectrum* and is equivalent to each of the $G(\infty)$ values at each time multiplied by the probability of that process. Note that here the temporal behaviour is expressed on a logarithmic scale as often we observe very broad responses. Of course, if our material is a viscoelastic liquid, then $G(0) = 0$ in Equations (9.32) and (9.33). Thus, for a viscoelastic liquid, such as the shower gel example shown in Figure 9.11, we can write the 'generalized Maxwell model' equations as follows:

$$G'(\omega) = \sum_{1}^{n} G(\infty)_n \frac{(\omega\tau_n)^2}{1 + (\omega\tau_n)^2}$$

$$G''(\omega) = \sum_{1}^{n} G(\infty)_n \frac{(\omega\tau_n)}{1 + (\omega\tau_n)^2} \qquad (9.34)$$

which in integral form are:

$$G'(\omega) = \int_{-\infty}^{+\infty} H \frac{(\omega\tau)^2}{1 + (\omega\tau)^2} d\ln \tau$$

$$G''(\omega) = \int_{-\infty}^{+\infty} H \frac{(\omega\tau)}{1 + (\omega\tau)^2} d\ln \tau \qquad (9.35)$$

At this point, we can see the similarity with the characterizing of materials by spectroscopic methods in the microwave, infrared, etc. However, we rarely see sharp, well separated peaks. So, although our relaxation spectrum is really the summary of our material behaviour, recording the maximum peak height(s) and the frequency at which it (they) occur does not provide enough information. This is because in nearly all cases we are looking at multiple processes which are located closely in time and we should note from Figure 9.10 how a single process has a curve which displays a 'half-width half-height' of ~ 1.5 orders of magnitude in frequency.

5.2 The Limiting Behaviour at High Frequencies

To estimate the limiting responses at high frequencies, that is, as $\omega \to \infty$, we should note that for a *viscoelastic liquid*:

$$G^*(\omega)_{\omega \to \infty} \to G(\infty) = \sum_1^n G(\infty)_n \text{ or } = \int_{-\infty}^{+\infty} H \mathrm{d} \ln \tau$$

$$\text{as } \frac{(\omega\tau)^2}{1 + (\omega\tau)^2} \to 1, \text{ and } \frac{(\omega\tau)}{1 + (\omega\tau)^2} \to 0 \tag{9.36}$$

The low-frequency limit is also of interest and in this case is expressed in terms of the limiting value of the complex viscosity, or the zero-shear viscosity, as follows:

$$\eta^*(\omega)_{\omega \to 0} = \frac{G^*(\omega)_{\omega \to 0}}{\omega} \to \eta(0) = \sum_1^n G(\infty)_n \tau_n \text{ or } = \int_{-\infty}^{+\infty} H\tau \mathrm{d} \ln \tau$$

$$\text{as } \frac{\omega\tau^2}{1 + (\omega\tau)^2} \to 0, \text{ and } \frac{(\tau)}{1 + (\omega\tau)^2} \to \tau \tag{9.37}$$

The modelling of the zero-shear viscosity has already been discussed for some colloidal dispersions and there are models available in the literature for polymers in solution [11]. The modelling of the high-frequency modulus will be discussed later in this chapter.

Carrying out the same exercise for a *viscoelastic solid* yields:

$$G(\infty) = G(0) + = \sum_1^n G(\infty)_n \text{ or } = G(0) + \int_{-\infty}^{+\infty} H \mathrm{d} \ln \tau \tag{9.38}$$

The low-frequency limit is:

$$\frac{G^*(\omega)_{\omega \to 0}}{\omega} \to \eta(0) = \frac{G(0)}{\omega \to 0} + \sum_{1}^{n} G(\infty)_n \tau_n \to \infty$$

$$\text{or } = \frac{G(0)}{\omega \to 0} + \int_{-\infty}^{+\infty} H\tau \mathrm{d} \ln \tau \to \infty \qquad (9.39)$$

which is, of course, the correct result for a *solid*! As an illustrative exercise, we will take the spectra used to fit the data shown in Figure 9.13 with and without the $G(0)$ term to show the frequency-dependence of the complex viscosity. It is quite clear from Figure 9.15, that without the $G(0)$ value in the four-component spectrum, the viscosity reaches a plateau value of $\sim 350\,\mathrm{Pa\,s}$ at frequencies below $10^{-2}\,\mathrm{rad\,s^{-1}}$. When there is a solid-like component, the viscosity increases steadily towards infinity with unit slope.

It is appropriate to consider the implications of this in terms of the product behaviour. When such a formulation has been prepared to be non-sedimenting, the solid-like response would appear to be desirable. However, the

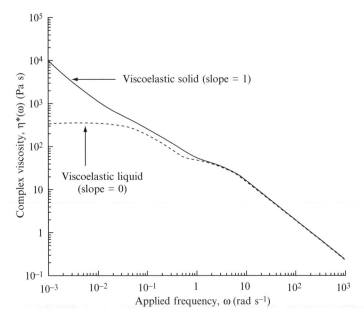

Figure 9.15. The complex viscosity calculated for the four-component viscoelastic spectrum used in Figure 9.13 and for a viscoelastic liquid from the same spectrum with the $G(0)$ value $= 0$.

plateau value of the low-shear viscosity of the fluid-like system is sufficiently high that, for abrasive particles such as calcium carbonate, if the radius of the largest particles or aggregates is < 2 or $3\,\mu m$ the sedimentation rate is so slow that it would take ~ 3 months for a 1 mm layer of clear fluid to form. This, however, may be acceptable if the length of time spent by the formulation in storage is usually less than this.

5.3   Stress Relaxation or the Step-Strain Experiment

This experiment is the mechanical analogy of the temperature jump and pressure jump experiments used to study chemical equilibria. In this case, we apply a small shear strain to the material as rapidly as possible and then follow the resulting stress as a function of time. A suitable starting point to describe the behaviour is to consider the response of a single Maxwell model, that is, a linear viscoelastic fluid with a single relaxation time. This is then the analogue of a first-order chemical reaction approach to equilibrium and we follow the stress from its initial value, σ_0, to 0 over time with a constant strain of γ_0. A curve such as that shown in Figure 9.16 will be obtained. We can express the rate of change of stress at any time as follows:

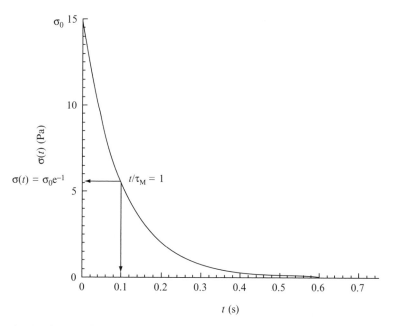

Figure 9.16.   Stress relaxation for a linear viscoelastic 'Maxwell 1' fluid with a characteristic time of 1 s after being subjected to a step strain of 5 %.

$$\frac{d\sigma(t)}{dt} = -k\sigma(t) \tag{9.40}$$

where k is the first-order rate constant for this relaxation and thus has units of s^{-1}. This is the reciprocal of the characteristic or Maxwell relaxation time, τ_M. We can write Equation (9.40) in terms of the stress relaxation modulus, $G(t)$, by dividing through by the value of the constant applied strain:

$$\frac{dG(t)}{G(t)} = -kdt \tag{9.41}$$

This equation is then integrated to give the value of the relaxation modulus at time t, as follows:

$$\int \frac{dG(t)}{G(t)} = -\frac{1}{\tau_M} \int dt \tag{9.42}$$

The limits for the integration are $t = 0$ and t, and so we have:

$$\ln\ G(t) - \ln\ G(t \to 0) = -\frac{t}{\tau_M} \tag{9.43}$$

that is:

$$G(t) = G(t \to 0)\exp\left(-\frac{t}{\tau_M}\right) \tag{9.44}$$

This can be generalized to include as many processes as are required to describe the curves:

$$G(t) = \sum_{i=1}^{i=n} G_i \exp\ (-t/\tau_i) \quad \text{(fluid)} \tag{9.45a}$$

$$G(t) = G(0) + \sum_{i=1}^{i=n} G_i \exp\ (-t/\tau_i) \quad \text{(solid)} \tag{9.45b}$$

Figure 9.17 shows the plots for the Maxwell liquid shown in Figure 9.16 (corresponding to the data found for the shower gel presented in Figure 9.11) and the four-component model used to describe the solid-like behaviour of the domestic cleaner in the oscillation experiment plotted in Figure 9.13. The integral forms of the stress relaxation equations for the generalized Maxwell models are as follows:

$$G(t) = \int_{-\infty}^{+\infty} H(\ln\ t)\exp\ (-t/\tau)d\ln\ \tau \quad \text{(fluid)} \tag{9.46a}$$

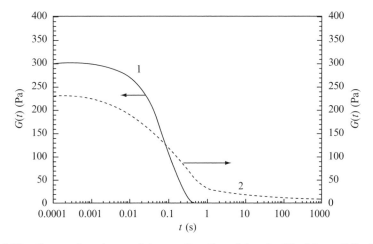

Figure 9.17. Stress relaxation modulus as a function of time for (1) a Maxwell liquid with $G = 300$ Pa and $\tau = 0.1$ s, and (2) a four-component solid with $G(0) = 10$ Pa, $G_1 = 20$ Pa, $G_2 = 145$ Pa and $G_3 = 55$ Pa, with $\tau_1 = 15$ s, $\tau_2 = 0.25$ s and $\tau_3 = 0.01$ s.

$$G(t) = G(0) + \int_{-\infty}^{+\infty} H(\ln t)\exp\,(-t/\tau)\mathrm{d}\ln\,\tau \quad \text{(solid)} \qquad (9.46b)$$

The stress relaxation experiment has some strong points in its favour. For example, it is much more rapid to carry out than applying a series of discrete oscillation frequencies, especially if very low frequencies (long timescales) are important. However, there are some other points that we need to keep in mind. The first is that rapid responses can be lost during the period taken to apply the step strain. Secondly, large strains are always applied during the loading of the instrument with the sample and these may not be applied evenly. It is therefore useful to apply an oscillating strain with an initially large but decreasing amplitude to remove any residual directional stress components, and then allow sufficient recovery time for structural 'rebuild' to take place prior to starting the experiment. This is also good practice with the oscillation experiment. Thirdly, we should keep in mind that the solution to a sum of exponential functions is difficult mathematically. This is known as an 'ill-conditioned' problem and will not give a unique solution. The simplest approximation to estimating the spectrum from the relaxation curve [11] is built into some equipment software and is as follows:

$$H(\ln t) \approx \frac{-\mathrm{d}G(t)}{\mathrm{d}\ln t} \qquad (9.47)$$

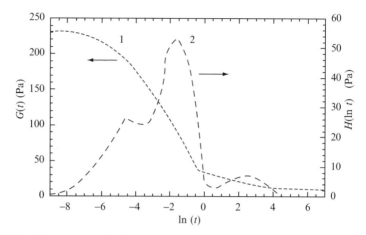

Figure 9.18. The stress relaxation modulus (1) and the relaxation spectrum (2) as a function of time: $\tau_1 = 15$ s; $\tau_2 = 0.25$ s; $\tau_3 = 0.01$ s.

The stress relaxation modulus calculated for the four-component model plotted in Figure 9.17 is re-plotted in Figure 9.18 as a function of ln t, along with the relaxation spectrum calculated from Equation (9.47). Peaks corresponding to the three relaxation times are clearly visible. The area under the spectrum, when added to the $G(0)$ value of 10 Pa, is equal to the value of $G(t \rightarrow 0) = G(\infty) = 230$ Pa (see Equation (9.38)). The individual relaxation times are resolved but only the value of G_1 can be readily estimated just by approximating the peak to a triangular shape of height 10 Pa and a width of 4, giving an area of 20 Pa. Estimation of the values of G_2 and G_3 would require a model of the peak shapes, as the two peaks overlap each other.

5.4 Creep Compliance

In this experiment, a step stress is applied and the compliance is followed as a function of time. With the modern range of controlled-stress instruments it is easier to apply the stress in very short times than in the step-strain experiment. The compliance, $J(t)$, (the measured strain per unit applied stress) is recorded as a function of time. Models such as those illustrated in Figures 9.8(d) and 9.8(f) are typical of those used to describe the response. Figure 9.8(d) shows a model of a solid with a retarded response – this is known as a *Kelvin* or *Voight* model. Springs or dashpot dampers can be added to provide an instantaneous elastic response or a fluid response, respectively. The *Burger* body shown in Figure 9.8(f) has both. The compliances of the individual elements in series are simply added to give the compliance of the whole model, as follows [12]:

$$J(t) = \frac{\gamma(t)}{\sigma_0} = \frac{1}{G}\left[1 - \exp\left(-\frac{t}{\tau_K}\right)\right] \quad \text{(Kelvin model – solid)} \quad (9.48)$$

$$J(t) = \frac{1}{G_2} + \frac{1}{G_1}\left[1 - \exp\left(\frac{t}{\tau_K}\right)\right] + \frac{t}{\eta_2} \quad \text{(Burger body – liquid)} \quad (9.49)$$

where $\tau_K = \eta_1/G_1$ and is known as the *Kelvin retardation time*. Note that here η_2 is the zero shear viscosity, $\eta(0)$, and G_2 is equal to $G(\infty)$ of the material described by the model.

In addition, we should note that $1/G(\infty) = J_g$, the 'glassy compliance' of the material. Curves calculated from Equations (9.48) and (9.49) are plotted in Figure 9.19 for both the creep and recovery responses after the application of a step stress of 1 Pa. Note that only the elastic components give recovery – the viscous compliance is permanent.

The limiting slope of the creep curve gives the zero-shear viscosity (that is $\eta(0) = \eta_2$ in Equation (9.49)) but there is often an experimental problem associated with its determination from this part of the data. We have to be sure that we are still working at small enough strains for which linear behaviour is occurring and quite large strains may be used before the experimental curve appears to have become linear. It is straightforward to see if too large a strain has occurred as the elastic recovery should be the same as the initial elastic strain. So, although it is tempting to leave the stress applied to the sample until the compliance response becomes linear, the likely result will be a very small elastic recovery. The inverse of the slope then is not the zero-shear

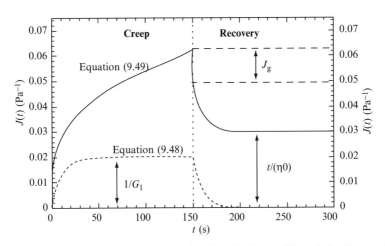

Figure 9.19. Creep and recovery curves for the Kelvin model and the Burger body calculated by using Equations (9.48) and (9.49): $G_1 = 50\,\text{Pa}$; $\tau_K = 10\,\text{s}$; $G_2 = J_g = 80$ Pa; $\eta_2 = \eta(0) = 5\,\text{kPa s}$.

viscosity because the structure has been markedly perturbed by the strain – here, strain melting has occurred. The solution is to apply the stress for a short enough time to ensure that only small strains are induced and then to analyse the recovery curve.

The response described by the constitutive equation for the Burger body (Equation (9.49)) has three elements. This may be generalized by adding a range of retardation times, either as a summation or in an integral form, as follows:

$$J(t) = J_g + \sum_{1}^{n} \left[1 - \exp\left(-\frac{t}{\tau_i} \right) \right] \quad \text{(solid)} \tag{9.50a}$$

$$J(t) = J_g + \sum_{1}^{n} \left[1 - \exp\left(-\frac{t}{\tau_i} \right) \right] + tJ_v \quad \text{(liquid)} \tag{9.50b}$$

where the steady state or viscous compliance is J_v ($= \eta(0)$). The integral forms of these equations are:

$$J(t) = J_g + \int_{-\infty}^{+\infty} L(\ln t) \left[1 - \exp\left(-\frac{t}{\tau} \right) \right] d\ln \tau \quad \text{(solid)} \tag{9.51a}$$

$$J(t) = J_g + tJ_v + \int_{-\infty}^{+\infty} L(\ln t) \left[1 - \exp\left(-\frac{t}{\tau} \right) \right] d\ln \tau \quad \text{(liquid)} \tag{9.51b}$$

where $L(\ln t)$ is the retardation spectrum of the material. This may also be determined from oscillating stress experiments carried out at different frequencies. We should note that for a linear viscoelastic material the complex compliance is as follows:

$$J^* = G^{*-1} \tag{9.52}$$

The dynamic compliances are given by:

$$J'(\omega) = \int_{-\infty}^{+\infty} L(\ln t) \frac{1}{1 + (\omega\tau)^2} d\ln \tau \tag{9.53a}$$

$$J''(\omega) = \int_{-\infty}^{+\infty} L(\ln t) \frac{\omega\tau}{1 + (\omega\tau)^2} d\ln \tau \tag{9.53b}$$

The manufacturers of controlled stress rheometers frequently output the data as dynamic moduli by using the following relationships:

$$G'(\omega) = \frac{J'(\omega)}{[J'(\omega)]^2 + (J''(\omega)]^2} \tag{9.54a}$$

$$G''(\omega) = \frac{J''(\omega)}{[J'(\omega)]^2 + (J''(\omega)]^2} \tag{9.54b}$$

However, it is important to note that the relationships expressed by Equations (9.54a) and (9.54b) are only true as long as the material is within the linear range of behaviour. As with all of the experiments described in this section, the first stage in the analysis of a new material should be to determine its linear range of response.

5.5 The Limiting Behaviour at Long or Short Times

The limiting responses at long or short times can be important in the end uses of colloidal materials and are often important as characterization parameters. For example, Equations (9.36) and (9.38) give the high-frequency limit to the storage modulus in terms of the integral of the whole relaxation spectrum for a liquid-like system and a solid-like system, respectively, whereas Equation (9.37) gives the zero-shear viscosity for the liquid-like system derived from the spectrum. The challenge is to predict these values from our knowledge of the colloidal particles. The zero-shear viscosity of colloidally stable and concentrated dispersions of charged spherical particles has already been discussed. The problem of the viscosity of long rods has much in common with that of semi-dilute polymer solutions [13] utilizing reptation dynamics. Some of the models have also been reviewed by Goodwin and Hughes [5].

Modelling of the high-frequency elastic response has not been discussed so far although in some cases it is relatively straightforward to carry out [5]. It is useful to think in terms of the family of material properties consisting of the internal energy density, the excess osmotic pressure (that is, just the contribution due to the interactions between the particles) and the high-frequency shear modulus of the dispersion [5]:

$$\frac{\bar{E}a^3}{k_B T} = \frac{9\varphi}{8\pi} + \frac{3}{2}\varphi \int_0^\infty r^2 g(r) \frac{V(r)}{k_B T} dr \tag{9.55a}$$

$$\frac{\Pi a^3}{k_B T} = \frac{3\varphi}{4\pi} - \frac{3\varphi^2}{8\pi a^3} \int_0^\infty r^3 g(r) \frac{d}{dr}\left(\frac{V(r)}{k_B T}\right) dr \tag{9.55b}$$

$$\frac{G(\infty)a^3}{k_B T} = \frac{3\varphi^2}{40\pi a^3} \int_0^\infty g(r)\frac{d}{dr}\left[r^4\frac{d}{dr}\left(\frac{V(r)}{k_B T}\right)\right] dr \qquad (9.55c)$$

In each of these equations, we are integrating the interaction over the dispersion structure whether this is the interaction energy, the interparticle force or the 'colloidal spring constant'. The particle size and volume fraction are readily determined experimentally and we can calculate the pair-potential. In most colloidal systems, the first nearest-neighbour interactions dominate the response and so the use of the pair-potential is often a reasonable approximation. The problem is then the pair-distribution function. In principle, this may be determined from the inversion of the structure factor obtained from scattering, but it may also be modelled by using the statistical mechanical techniques for molecular liquids. Comparison of the values calculated from Equation (9.55c) with those obtained from experiments using models for spherical particles [14], with a short-range attractive minimum in the pair-potential, show excellent agreement. These dispersions contained particles with a weak attractive minimum between the particles. It is interesting to note that the longer the range of the interaction, then the lower the volume fraction at which elastic responses appear. Fluid-like systems can be found with the longer-range attractions, while rigid solid phases are produced when the interaction is of very short range when compared to the particle radius. We should also note that we are dealing here with non-equilibrium structures. Just the process of mixing the systems or placing them into a rheometer cell for measurement tends to homogenize the structures, which is one of the reasons that methods developed for liquid-state calculations can be so successful for systems which are clearly solids in that yield-behaviour $G(0)$ values can be seen.

The situation can be simplified when there is a long-range repulsion between the particles with a narrow size distribution. In this case, the local structure becomes ordered which simplifies the structure factor. The separation is the same for all of the nearest neighbours and the structure has face-centred-cubic (fcc) symmetry. In some cases, long-range order can be produced, that is, we have a 'colloidal crystal'. However, it is much more usual for the order to be short range and we should think in terms of a 'colloidal glass'. The regularity of the spatial aspects of the structure enables Equations (9.55a–9.55c) to be written more simply for a spherical cell with a coordination number of z and a mean centre-to-centre particle separation as:

$$\frac{\bar{E}a^3}{k_B T} = \frac{9\varphi}{8\pi} + \frac{3\varphi z}{8\pi}\frac{V(R)}{k_B T}, \text{ with } R = 2a\left(\frac{\varphi_m}{\varphi}\right)^{1/3} \qquad (9.56a)$$

$$\frac{\Pi a^3}{k_B T} = \frac{3\varphi}{4\pi} - \frac{\varphi z}{8\pi}\frac{d}{dR}\left(\frac{V(R)}{k_B T}\right) \qquad (9.56b)$$

$$\frac{G(\infty)a^3}{k_BT} = \frac{\varphi z R^2}{40\pi} \frac{d^2}{dR^2}\left(\frac{V(R)}{k_BT}\right) \tag{9.56c}$$

A similar result to Equation (9.56c) can be derived by using a 'zero-Kelvin lattice model' [5] (zero-Kelvin, so there is no relaxation). With a slightly different spatial averaging process, the result was:

$$\frac{G(\infty)a^3}{k_BT} = \frac{3\varphi z R^2}{256\pi} \frac{d^2}{dR^2}\left(\frac{V(R)}{k_BT}\right) \tag{9.56d}$$

The effectiveness of Equation (9.56d) in accurately describing the elastic response has been demonstrated by using a range of charge-stabilized monodisperse particles. Again, we should note that the structures are non-equilibrium ones due to the preparation and handling techniques employed and are similar to colloidal glasses. (The equilibrium structure would be in the form of colloidal crystals.) These can be observed with small particles just above the phase transition. The forces are so weak that the long-range structure is readily broken down on very gentle stirring, although the short-range structure is maintained.

Polymers are often used to produce weak gels so that sedimentation is controlled at the same time as high-shear viscosity is maintained at a chosen level. With polymer gels, we are not usually concerned with the high-frequency limit of the storage modulus but the plateau or network modulus. This is observed over a long range of timescales [11, 13]. The faster processes which result in the $G(\infty)$ are characterized by the relaxation times associated with the motion of one or two repeat units on the chain. The much slower cooperative motion of reptation governs the other end of the timescale observed for simple homopolymers. With cross-linked polymers, the lifetime of the links is much longer than that of an entanglement. These cross-links can be due to covalent bonds, ion–chain interactions, particle chain interactions or hydrophobe self-assembly. The simplest treatment is to model the system as a swollen elastomeric network so that the network modulus, G_N, is simply proportional to the number of links contributing to the network, N_L:

$$G_N = N_L k_B T \tag{9.57}$$

There are models available to calculate N_L for the various types of polymer network [5]. We should note that N_L is the number density of *effective* links and so requires some statistical modelling in most cases.

5.6 Processing Effects

The key factors that we need to aid our understanding of the behaviour of concentrated dispersions are the interactions between particles and the structure of the system. It is usually possible to calculate the pair-potential even if the potential of mean force may be more difficult to estimate. The structure is the major problem. Whenever we measure our materials, they will have been subjected to large deformations and the effects of these are frequently important. For example, we can set up a model experiment to study diffusion-limited aggregation and start to observe fractal clusters being formed. As soon as the system is transferred to a container or an instrument cell, it is subjected to large deformations and we will then be studying a *processed* system. This may work in our favour in many instances but it may also lead to long-term changes on storage. To clarify these points, we will take one or two examples.

In some instances, concentrated systems which are coagulated are used. Clay slurries for slip-casting, or at higher concentrations for ceramic materials, are examples of these. If a dispersion is caused to coagulate and then pumped or spread, the shear processing will do two things. It will tend to densify the local structure and then break this up into flow units made up of many particles. Spherical polymer particles coagulated with electrolyte and sheared in a controlled fashion produce monodisperse dense spherical clusters [5]. The interacting units then become the large clusters and so any model based on the summation of all of the interactions between all of the single particles in the system would fail as a large overestimate of the resistance to continuous motion.

The models used above for the behaviour of concentrated dispersions of uniform-sized particle, which are either structured due to strong repulsive interactions or weak attractions, are based on a structural uniformity which is based on short-range structures. In both cases, the lowest energy state that we can visualize would be large crystals rather than the glassy or liquid-like structures that we use. These models work in a satisfactory manner because the intensive mixing that is a part of the process of loading the systems into containers or measuring equipment produces these metastable structures. The relaxation times are often very long – days, weeks or even years. If the weakly attractive system is explored a little further we can sometimes see unexpected phenomena, such as delayed settling. What occurs is that the shearing forces are large enough to pull the individual particles apart and distribute them evenly. On the cessation of shearing, the particles form a uniform liquid-like structure with a coordination number that is a function of the volume fraction. The motion of the individual particles allows slow local densification to occur as, for example, twelve nearest neighbours is a lower energy state than eight. The local densification leads to other regions with fewer particles. As such weaknesses grow, a point can be reached where the structure collapses and the liquid is expelled from the voids. The collapse can be quite rapid once the system reaches a critical state, although it can take a long time to reach that condition.

Mixtures formed with different particle sizes and/or densities can show structural changes as the result of low shear forces. For example, the

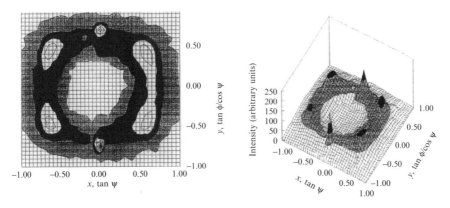

Figure 9.20. Hexagonal spot patterns obtained from diffraction in an optical rheometer for a sample under oscillatory shear [15]. Reproduced by permission of H. Sakabe, 'The Structure and Rheology of Strongly Interacting Suspensions', *PhD Thesis*, University of Bristol, Bristol, UK, 1995.

structure of some systems may become sufficiently broken down, i.e. the viscosity has been lowered, by the low frequency, although large-amplitude deformations produced when transporting bulk shipments may mean that large or dense particles can separate out. Simple jar tests in the laboratory may not show this.

Another interesting example concerning the effect of low shear is that on concentrated systems formed from stable particles. Oscillations can enhance the structuring rate and large amplitudes are more effective than small. Figure 9.20 shows the optical diffraction patterns produced in an optical rheometer [15]. Systems of bimodal particles can be induced to crystallize by slow rotation if the particle size ratios and concentrations are carefully chosen [16].

REFERENCES

1. R. J. Hunter, *Foundations of Colloid Science*, Vol. II, Oxford University Press, Oxford, UK, 1989.
2. R. J. Baxter, *Aust. J. Phys.*, **21**, 563 (1968).
3. R. H. Ottewill, 'Concentrated Dispersions', in *Colloidal Dispersions*, J. W. Goodwin (Ed.), The Royal Society of Chemistry, London, 1982, pp. 197–217.
4. J. W. Goodwin and J. Mercer-Chalmers, 'Flow Induced Aggregation of Colloidal Particles', in *Modern Aspects of Colloidal Dispersions*, R. H. Ottewill and A. Rennie (Eds), Kluwer Academic Publishers, Dordrecht, The Netherlands, 1998, pp. 61–75.
5. J. W. Goodwin and R. W. Hughes, *Rheology for Chemists – An Introduction*, The Royal Society of Chemistry, Cambridge, UK, 2000.

6. M. Krieger and T. J. Dougherty, *Trans. Soc. Rheol.*, **3**, 137 (1959).
7. R. Ball and P. Richmond, *Phys. Chem. Liq.*, **8**, 99 (1980).
8. M. E. Woods and I. M. Krieger, *J. Colloid Interface Sci.*, **34**, 417 (1970).
9. T. Ohtsuki, *Physica A*, **108**, 441 (1981).
10. W. B. Russel, D. A. Saville and W. R. Schowalter, *Colloidal Dispersions*, Cambridge University Press, Cambridge, UK, 1989.
11. J. D. Ferry, *Viscoelastic Properties of Polymers*, 3rd Edn, Wiley-Interscience, New York, 1980.
12. R. I. Tanner, *Engineering Rheology*, Oxford University Press, Oxford, UK, 1985.
13. M. Doi and S. F. Edwards, *The Theory of Polymer Dynamics*, Oxford University Press, Oxford, UK, 1986.
14. J. W. Goodwin, R. W. Hughes, S. J. Partridge and C. F. Zukoski, *J. Chem. Phys.*, **85**, 559 (1986).
15. H. Sakabe, 'The Structure and Rheology of Strongly Interacting Suspensions', *PhD Thesis*, University of Bristol, Bristol, UK, 1995.
16. G. D. W. Johnson, R. H. Ottewill and A. Rennie, 'Characterisation of Particle Packing', in *Modern Aspects of Colloidal Dispersions*, R. H. Ottewill and A. Rennie (Eds), Kluwer Academic Publishers, Dordrecht, The Netherlands, 1998, pp. 89–100.

Index

Colloids and Interfaces with Surfactants and Polymers – An Introduction   J. W. Goodwin
© 2004 John Wiley & Sons, Ltd   ISBN: 0-470-84142-7 (HB)   ISBN: 0-470-84143-5 (PB)